LASERS AND ELECTRO-OPTICS RESEARCH AND TECHNOLOGY

ORGANIC LIGHT-EMITTING DIODES (OLED)

MATERIALS, TECHNOLOGY AND ADVANTAGES

LASERS AND ELECTRO-OPTICS RESEARCH AND TECHNOLOGY

Additional books in this series can be found on Nova's website
under the Series tab.

Additional e-books in this series can be found on Nova's website
under the e-book tab.

LASERS AND ELECTRO-OPTICS RESEARCH AND TECHNOLOGY

ORGANIC LIGHT-EMITTING DIODES (OLED)

MATERIALS, TECHNOLOGY AND ADVANTAGES

DOUGLAS RIVERA
EDITOR

nova publishers
New York

Library of Congress Cataloging-in-Publication Data

Names: Rivera, Douglas.
Title: Organic light-emitting diodes (OLED) : materials, technology and advantages / editor, Douglas Rivera.
Description: Hauppauge, New York : Nova Science Publishers Inc., [2015] | Includes index.
Identifiers: LCCN 2015038878 | ISBN 9781634840019 (hardcover)
Subjects: LCSH: Light emitting diodes. | Organic semiconductors.
Classification: LCC TK7871.89.L53 O7338 2015 | DDC 621.3815/22--dc23 LC record available at http://lccn.loc.gov/2015038878

Published by Nova Science Publishers, Inc. † New York

CONTENTS

PREFACE

Organic light-emitting diodes (OLEDs) are considered as the most promising technology for next generation display and solid-state lighting for their advantages such as surface emitting, ease for large area manufacturing, viability for flexible and transparent applications, low energy consumption and potential to be low-cost. Chapter one of this book discusses tandem OLEDs, which are technologically interesting because not only can the luminance and current efficiency be improved linearly with the number of electroluminescent (EL) units in the tandem OLED, but also leakage current and breakdown of the electric field can be avoided due to the higher luminance at a low current density and the thicker organic films, resulting in a long lifetime. Chapter two mainly deals with the low cost synthesis and characterization of amino, amino-chloro, dichloro and trichloro substituted derivatives of diphenyl quinoline (DPQ). Chapter three aims at reporting the last strategies that have been developed to address the issue of roll-off.

Chapter 1 – Tandem OLEDs are technologically interesting because not only can the luminance and current efficiency be improved linearly with the number of electroluminescent (EL) units in the tandem OLED, but also leakage current and breakdown of the electric field can be avoided due to the higher luminance at a low current density and the thicker organic films, resulting in a long lifetime. Importantly, the state-of-the-art tandem OLEDs are very easy to vertically stack either individual red, green, and blue emission units or multiple white emission units in series via charge generation layers (CGLs) to achieve white emission. In tandem OLEDs, the CGL plays a critical role in the realization of high-performance tandem OLEDs since it functions as both an internal anode and a cathode to generate intrinsic charge carriers and to facilitate opposite electrons and holes being injected into the adjacent sub-OLED units. Although amounts of CGLs have been developed to use tandem OLEDs, the problems still remain including complicated processing and limited material combination. More importantly, since the driving voltage consumed by conventional tandem devices scales linearly with the number of electroluminescent units, the resulting power consumption would be the same for both the single-unit and tandem OLEDs to obtain the same luminescence, this means that the power efficiency cannot be increased for such tandem devices when using general CGLs. Therefore, developing new CGL structures have become important research topic in this field. In this chapter, the authors will review the progress of tandem OLEDs based on different CGLs from structure design to their working principles. The more discussion will be focused on the organic semiconductor heterojunctions as CGLs in tandem OLEDs. As the authors see, besides the luminance and current efficiency are double enhancement, the power efficiency of resulting tandem OLEDs based on organic heterojunctions as CGLs is also greatly improved. Here, the authors will give detailed discussions on the structure design of organic heterojunctions as CGLs and the mechanism of their charge generation.

Chapter 2 – Since the preceding decade, OLEDs have become globally recognized exceptional area light source, which has enormous potential for the recent display applications and can overlay a novel path to create white light. In this regard, solid state lighting (SSL) - the new-fangled green technology has the prospective to go beyond the energy efficiencies of incandescent and fluorescent lighting and make the lighting eco - friendly and energy efficient. Such lighting can be achieved from light emitting devices based on either sublimely organics or polymeric materials and hence considerable attention has been grabbed in the scientific as well as in technological fields. Among the three primary RGB colours, the synthetic protocols and fabrication methods of green and red phosphors meet the necessary requirements. Conversely, the design and fabrication of blue phosphors and devices is still an ongoing challenge and hence research speaks volumes about the key challenges in attaining energy efficient blue light emitting organic emissive materials.

This chapter mainly deals with the low cost synthesis and characterization of amino, amino-chloro, dichloro and trichloro substituted derivatives of diphenyl quinoline (DPQ) by Friedlander condensation. The preparation of blended thin films with the synthesized complexes and binding polymers such as poly (methyl methacrylate) (PMMA) and polystyrene (PS) at different weight % are also illustrated. One of the main reasons for their synthesis and investigation of the present new conjugated polymeric compounds is to explore the molecular engineering approach as a means of the structural basis for the optimization of the desired solid-state properties. A single technique can not provide complete analysis of a solid and hence variety of instrumentation techniques were used for characterization of the synthesized materials. They include Fourier Transform Infrared Spectrometer (FTIR) and the X-ray diffraction (XRD) method, which are used to probe the structural properties. Spectrophotometer and spectroflurometer are used to study the optical properties by means of UV-Vis absorption and photoluminescence (PL) spectra. Thermal properties of the materials are analyzed by Thermo gravimetric Analysis (TGA) and Differential Thermal Analysis (DTA). The results reveal that the starting materials are affordable for the chemical synthesis of DPQ derivatives; they are thermally stable and offer high intensity light emission in blue region of electromagnetic spectrum. Hence, these materials can be considered as an ultimate substitute for the existing blue organic light-emitting materials, generally employed for OLEDs, display devices and many more applications in this photonic age.

Chapter 3 – Since the early work of Tang and Van Slyke, Organic Light-Emitting Devices (OLEDs) attract broad attention of researchers due to their potential uses in flat panel displays and solid-state lighting. Some OLEDs proponents even believe that white OLEDs (WOLEDs) will soon replace the existing fluorescent and filament lamps for general lighting application. At present, OLEDs still suffer from an insufficient lifetime and the transfer of this technology to the mass market is another story. Doping the emissive layer with fluorescent or phosphorescent dyes is widely practiced for devices fabrication. To reach high efficiencies, a key requirement is that all electrically generated excitons must be harvested for light emission. However, even if emitters and especially metal complexes are usually doped into suitable host materials, quantum efficiency of phosphorescent OLEDs still tends to decrease at high current density because of competitive factors such as concentration quenching and triplet-triplet annihilation. The efficiency roll-off i.e., the reduction of the quantum efficiency with increasing the current density is nearly general and constitutes the most significant problem facing OLEDs. This chapter aims at reporting the last strategies that have been developed to address this issue. Especially, a systematic understanding of the structure–property–performance relationships is proposed.

In: Organic Light-Emitting Diodes (OLED)
Editor: Douglas Rivera
ISBN: 978-1-63484-001-9

Chapter 1

TANDEM ORGANIC LIGHT EMITTING DIODES

Dongge Ma
State Key Laboratory of Polymer Physics and Chemistry, Changchun Institute of Applied Chemistry, University of Chinese Academy of Sciences, Changchun, State Key Laboratory of Luminescent Materials and Devices, South China University of Technology, Guangzhou, P. R. China

ABSTRACT

Tandem OLEDs are technologically interesting because not only can the luminance and current efficiency be improved linearly with the number of electroluminescent (EL) units in the tandem OLED, but also leakage current and breakdown of the electric field can be avoided due to the higher luminance at a low current density and the thicker organic films, resulting in a long lifetime. Importantly, the state-of-the-art tandem OLEDs are very easy to vertically stack either individual red, green, and blue emission units or multiple white emission units in series via charge generation layers (CGLs) to achieve white emission. In tandem OLEDs, the CGL plays a critical role in the realization of high-performance tandem OLEDs since it functions as both an internal anode and a cathode to generate intrinsic charge carriers and to facilitate opposite electrons and holes being injected into the adjacent sub-OLED units. Although amounts of CGLs have been developed to use tandem OLEDs, the problems still remain including complicated processing and limited material combination. More importantly, since the driving voltage consumed by conventional tandem devices scales linearly with the number of electroluminescent units, the resulting power consumption would be the same for both the single-unit and tandem OLEDs to obtain the same luminescence, this means that the power efficiency cannot be increased for such tandem devices when using general CGLs. Therefore, developing new CGL structures have become important research topic in this field. In this chapter, we will review the progress of tandem OLEDs based on different CGLs from structure design to their working principles. The more discussion will be focused on the organic semiconductor heterojunctions as CGLs in tandem OLEDs. As we see, besides the luminance and current efficiency are double enhancement, the power efficiency of resulting tandem OLEDs based on organic heterojunctions as CGLs is also greatly improved. Here, we will give detailed discussions on the structure design of organic heterojunctions as CGLs and the mechanism of their charge generation.

1. INTRODUCTION

Organic light-emitting diodes (OLEDs) are considered as the most promising technology for next generation display and solid-state lighting for their advantages such as surface emitting, ease for large area manufacturing, viability for flexible and transparent applications, low energy consumption and potential to be low-cost [1-5]. Since the breakthrough by Kodak in 1987, considerable efforts have been made to improve both their efficiency and lifetime in order to meet the requirements for mass production [6].

OLEDs are actually current driving device, with increasing current density through, the luminance would rise while operational lifetime would reduce [7-9]. Since tandem OLEDs (TOLEDs) can obtain the same luminance under several folds of lower current density (depend on the number of EL units), its lifetime can be significantly lengthened.

Furthermore, the tandem structure provides the feasibility that EL units of different colors can be vertically stacked together for color tuning and white emission [10-12]. Duing to the unique advantages over traditonal single unit OLEDs, TOLEDs have been attracting extensive research interests ever since their first appearance in 2003 [13].

A typical TOLED is fabricated by vertically connecting several individual electroluminescent (EL) units together in series via interconnectors called charge generation layers (CGLs), with the entire device driven by a single power source (Figure 1) [14-17]. When certain voltage is applied on the electrodes, each EL lights up individually under the same current that flows through the whole device. From the viewpoint of the simplest terms, the current efficiency and the external quantum efficiency of the tandem devices would be the sum of each EL unit; while the power efficiency might be inferior to the single EL unit ones, considering the extra voltage drop across the interconnector and the interfaces it brings in.

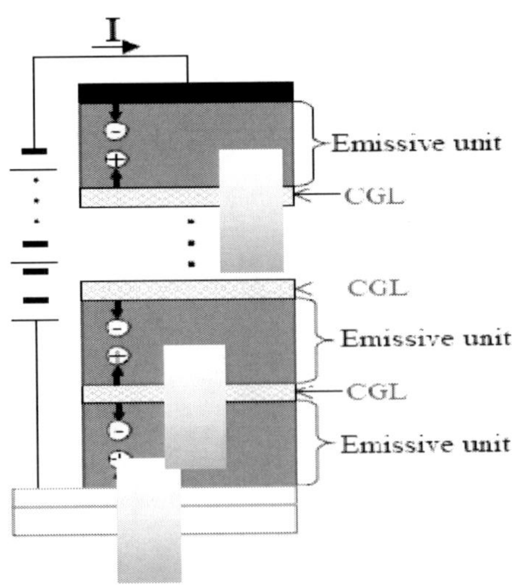

Figure 1. Structure and working processes of tandem OLEDs.

However, this shortcoming can be overcome as we will discuss in the article latter.

Understandably, rational interconnector selection and design are crucial to the performance of TOLEDs. A good CGL must meet several requirements: efficient hole-electron generation, minimal energy barrier for charge injection, high conductivity, transparent in visible spectral range, high operational stability and easy for deposition.

In fact, the researchers are working hard towards this goal, and most published papers on TOLEDs are focusing on interconnector design and its influence on device performance. Various types of interconnectors have been put forward, of which the most widely used structures include: metal-oxide, n-doped/p-doped organic bilayer and organic heterojunction (OHJ) [15-17].

2. Tandem OLEDs Based on Transition Metal-Oxide CGLs

The first TOLED reported by Kido et al. used ITO or V_2O_5 as interconnector between two EL units. Ever since, various interconnectors employing transition metal-oxides (TMOs) have been reported, of which MoO_3 and WO_3 are frequently used [15, 18-29]. The typical TMO type interconnector structure is to insert a thin layer of TMO between the hole-transporting layer (HTL) of one sub-EL unit and the electron-transporting layer (ETL) of another sub-EL unit.

For a long time, TMOs like MoO_3 has been considered as a p-type material which leads the misinterpretation of charges to generate at the interface of TMO and ETL [30]. In 2009, Kahn et al. found that TMOs actually have very low-lying LUMOs, 6.7 eV for MoO_3, for example [31]. Now it is generally considered that the charges are generated at the interface of TMO and adjacent hole transporting material (HTM) [19, 20]. For a typical TMO/HTM system, The TMO serves as a strong electron acceptor, when the electrons are drawn from the HTM HOMO to the LUMO of TMO, the holes would be left at the HTM HOMO side, thus electrons and holes are generated in pairs. Afterwards, the generated electrons and holes would separate and transport in opposite directions under certain bias, and are injected to adjacent ETL and HTL, respectively. This process is confirmed by several groups by the energy level alignment obtained from the ultraviolet photoemission spectroscopy (UPS) measurements [19, 20, 24, 25]. Among them, the results reported by Tang et al. are shown in Figure 2. The HTL and ETL interfaces without MoO_3 shows flat vacuum level alignment (Figure 2c and 2d), and no charge generation process is observed from additional C-V test [25]. The large energy level offset between NPB's HOMO and CsN_3:BPhen's (or BPhen's) LUMO indicates that the charge generation process is hard to occur at that interface unless higher driving voltage is applied. In contrast, the insertion of MoO_3 between HTL and ETL can induce large vacuum level bending, which is crucial for the electron-hole generation and injection (Figure 2a and 2b). Apart from the charge generation mechanism previously mentioned, the authors also suggest that the electron transfer in MoO_3 may happen spontaneously via thermal diffusion from various defect states to the conduction band; then under bias, the left holes in these gap states (Figure 2a and 2b) will tunnel into the HOMO of NPB, and the electrons in the conduction tunnel into the LUMO of BPhen.

Since TMOs have rather deep LUMOs, which is lower than the HOMOs of most common HTMs, the charge generation and the hole injection are relatively easy.

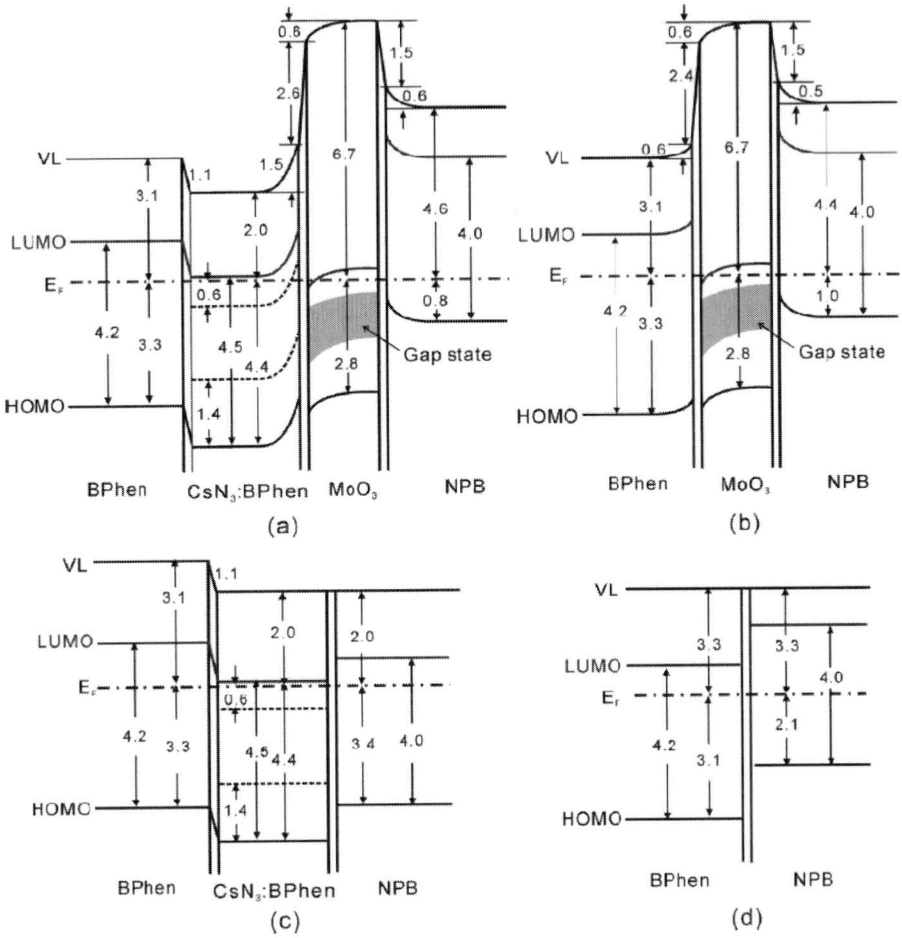

Reproduced from reference [25].

Figure 2. Schematic energy level diagrams: a) BPhen/CsN_3:BPhen/MoO_3/NPB, b) BPhen/MoO_3/NPB, c) BPhen/CsN_3:BPhen/NPB, and d) BPhen/NPB interfaces on ITO substrate, respectively. All the values shown are in the unit of eV.

However, the electron transporting materials (ETMs) in OLEDs usually have their LUMOs around 3.0 eV, therefore the electron injection from LUMO of TMO into the adjacent ETL is hard unless the vacuum level bending is involved (like in Figure 2). So the design of proper TMO/ETL interface is critical to the performance of TMOs based tandem devices. In Table 1, some of the reported TMO/ETL interfaces and their device performances are given. The vacuum level bending at the interface of intrinsic ETM and TMO is not very efficient, for example, the vacuum level is 3.1 eV above the Fermi level at the intrinsic BPhen side (Figure 2b), but only 2.0 eV above Fermi level if a n-doped ETL is added into BPhen (Figure 2a). Actually, the n-doped ETL is a commonly used solution to increase band bending at ETL/TMO interface. As seen in Table 1, most of the reported devices' ETL is n-doped, such as Mg, Li, Cs and their compounds are mostly used dopants. In comparison ([22] in Table 1), if the ETL is intrinsic, the device performance would be obviously inferior to the n-doped one.

Table 1. Comparison of the reported TMO based tandem devices

TMO/ETL interface	emitter	Tandem VS Single (max value given in References for two unit device)		References
		PE(lm/W)	CE(Cd/A)	
MoO_3/$CsCO_3$:BCP	DPAVBi	6.15/5.05	17.1/9.3	[18]
WO_3/Li:Alq	C545T	5.67/4.71	23.1/9.9	[15]
MoO_3/CsN_3:BPhen	Ir $(ppy)_3$	NA/57	84/41	[21]
MoO_3/$CsCO_3$:BPhen	C545T	4.19/3.84	22.16/10.7	[24]
MoO_3/Mg:Bphen	Ir $(ppy)_3$	NA/NA	67.9/33.2	[29]
MoO_3/LiF:BPhen MoO_3/Al/LiF:BPhen	Alq	1.2/2.0 2.5/2.0	7.0/3.5 8.8/3.5	[23]
MoO_3/Al/Liq/B3PyPB MoO_3/Al/Liq/Liq:B3PyPB	FIrpic	33.6/56.5 40.8/56.5	78.2/51.6 90.0/51.6	[22]

Reproduced from reference [22].

Another popular method to increase band bending at ETL/TMO interface is adding a thin layer of metal (about 1 nm) Al or Ag in between. From [23] in Table 1, we can see that the addition of a thin layer Al can improve the device performance significantly, especially in power efficiency, because the voltage drop at the electron injection interface is reduced. The mechanism of enhancing the charge generation by inserted metal is thought to be that the metal cluster can induce more gap state to assist charge tunneling [32, 33].

As we know, TMOs always have high evaporation temperature, which is one main drawback of TMOs based CGLs [15]. Moreover, there are also some concerns about the operating stability of TMO CGLs. In fact, the TMO interconnectors' stability is in close relation to the n-dopant in adjacent ETL, as reported by Tang et al. [26].

The authors prepared several tandem devices with MoO_3 as CGL, and only vary the n-dopants in ETL, the results exhibit rather significant difference in the EL spectra, as shown in Figure 3. The tandem devices use a green emitter in one EL unit and a red emitter in another EL unit. Tested for a period of time, the device with Mg:Alq/MoO_3 interface shows a synchronous decrease of both spectral intensity, but the devices with $CsCO_3$:Alq/MoO_3 and CsN_3:Alq/MoO_3 interfaces show a fast decrease of green spectral, which means that some degradations are taking place in the green EL unit. The author then exchanged the place of green and red EL units, finding that the red EL unit is degrading faster this time, the influence of electrodes are also excluded with other experiments.

All the phenomena points out that the degradation may happen at the interface of ETL/ MoO_3 interface. The degradation is attributed to the electric field induced migration of cesium ions towards ETL/MoO_3 interface, which will further lead to the reaction of cesium cations with MoO_3 to form complexes. Considering Li and Cs are the most frequently used n-dopant in OLEDs, the degradation problem of TMOs based CGLs might be common in tandem OLEDs based on metal-oxide CGLs.

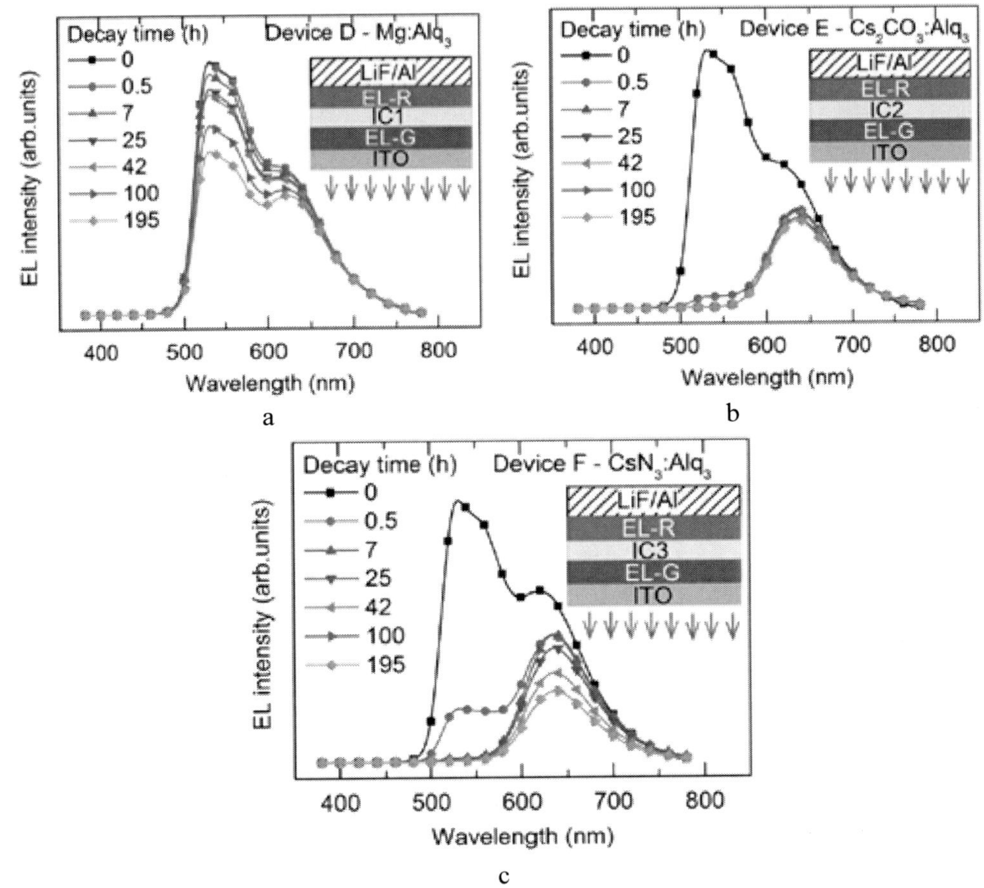

Reproduced from reference [26].

Figure 3. Time dependence of the emission characteristics of TOLEDs using various CGLs. EL spectra of devices (a) D, (b) E and (c) F versus operational time at a current density of 5 mA/cm^2. The insets show the device structures.

3. Tandem OLEDs Based on N-Doped/P-Doped Organic CGLs

The first CGL interconnector employing a n-doped organic layer and a p-doped organic layer forming bilayer n/p structure was reported in 2004 by Liao et al. [34]. They used Li:Alq/ $FeCl_3$:NPB as CGL to achieve a TOLED with current efficiency of 81.7 cd/A for two-unit device and 136.3 cd/A for three-unit device using $Ir(ppy)_3$ as emitter, which are about two and three times higher than single unit one, proving the validity of the n/p CGL. This type CGL is considered to offer several advantages over the TMOs, including excellent optical and electrical properties, as well as the ease of fabrication by thermal evaporation [34, 35]. A lot of other works using n/p type CGL interconnectors have been published in the last decades, among them, a Lewis acid or low LUMO materials such as F_4-TCNQ are always used as acceptor in the p-doped layer and alkali metals are used as donor in the n-doped layer [10, 16, 24, 34-40].

Similar to the TMO CGLs, the performance of n-doped/p-doped CGLs is closely related to the energy level alignment at the n/p interface. Due to the preserve of a large space-charge density in the doped systems, a significant energy shift at the interface can arise, which is essential to the charge generation of the CGLs [37, 38, 40, 41]. The popular understanding of the charge generation processes in the n-doped/p-doped systems involves charge tunneling.

In a paper published by Kröger et al. at 2007, a model for the charge generation was presented as direct tunneling from the HOMO states of the HTL to the LUMO states of the ETL [37]. The authors proved the model by conducting I-V test and Kelvin probe measurements. In this article, a series of devices with different interface structures are prepared to study the influence of doping on the I-V characteristics, which were shown in Figure 4b. The foreword profile of the I-V plot is considered to be a result of recombination by the injected carriers, while the reverse profile can be understood as the result of charge generation inside the devices.

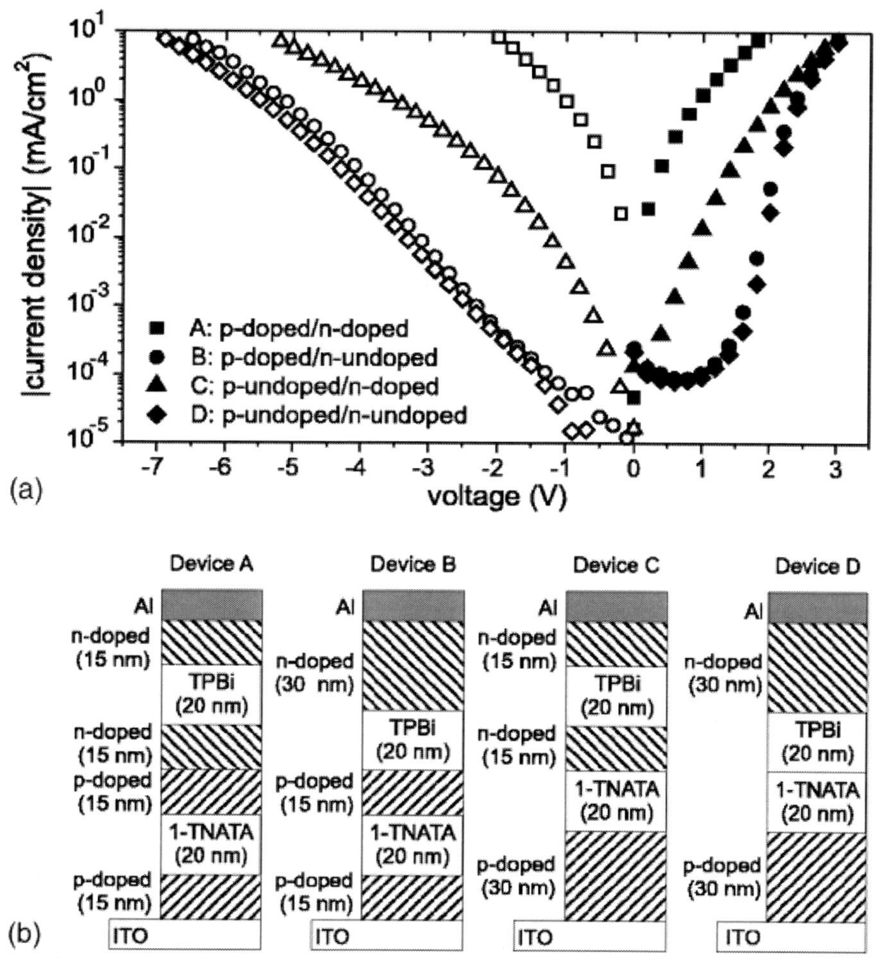

Reproduced from reference [37].

Figure 4. a) I-V characteristics of devices with different interface structures. b) Schematic diagrams of devices with different interface structures.

The carrier injection of the devices is very efficient due to the Ohmic contact, so the symmetry of the I-V characteristics of CGL structures can be taken as an indicator for the efficiency of charge generation at the n/p interface. It is found that both n-doping and p-doping are important for the generation of charges, only the interface with n-doping/p-doping structure shows the best symmetry, thus efficient charge generation, as shown in Figure 4a. Then the authors also tested the low temperature I-V characteristics of the devices with n-doped/p-doped interface, both foreword and reverse bias I-V curves are found to be independent of temperature. Since the foreword charge injection of such Ohmic contact device is usually interpreted as Fowler-Nordheim tunneling, it is natural to assume that the charge generation under reverse bias should also be tunneling.

The assumption is supported by the Kelvin probe measurements which show a rather thin charge depletion layer at the n/p interface, as well as very close to the p-doped material HOMO and n-doped material LUMO level alignment due to the vacuum level bending, suggesting that the electron tunneling would be very easy in this case, As shown in Figure 5a. In contrast, the undoped interface exhibits rather thick charge depletion layer (shown in Figure 5b), which will make the charge generation via tunneling difficult, therefore the necessity of doping is well explained. A schematic diagram is given in Figure 6, depicts the energy-level alignment and compares the involved physical processes for foreword and reverse bias. The similar tunneling charge generation mechanisms in n-doped/p-doped CGLs have also been suggested by other groups [38, 40].

From the above, we can see that the charge generation efficiency of the n-doped/p-doped CGLs mainly depend on the energy level alignment at the n/p interface, therefore sometimes the interface modifications are employed to increase the vacuum level shifting, for instance, a thin layer of Al or MoO_3, so as to improve the device performance [24, 35, 39, 40, 42].

The major drawback of n-doped/p-doped CGLs is also about the stability, the mutual diffusion of n-dopant and p-dopant will cause the variation of interface energy level alignment, thus would lead to a significant efficiency degradation [15, 38].

4. Tandem OLEDs Based on Organic Heterojunction CGLs

Due to the obvious drawbacks of TMO and n-doped/p-doped type CGLs mentioned above, and the power efficiencies of tandem devices based on them are far from satisfactory, therefore, the design of other new CGLs is very necessary. So far, organic heterojunction type CGLs seem to be the best choice for the architecting of high efficiency TOLEDs.

The advantages of organic heterojunction CGLs include non-doping thus more stable, easy for deposition, high optical transmission and efficient charge generation because of the good electrical properties [14, 15, 17, 43].

The first report about organic heterojunctions as CGLs was given by Lai et al. at 2007, they used copper hexadecafluorophthalocyanine ($F_{16}CuPc$)/copperphthalocyanine (CuPc) organic heterojunction as CGL to successfully achieve a TOLED with double current efficiency of the single unit device [15].

However, the operating voltage of the device increased significantly (more than triple times) due to the large injection barrier at $F_{16}CuPc$ and ETL interface, the resulting power efficiency is unsatisfactory.

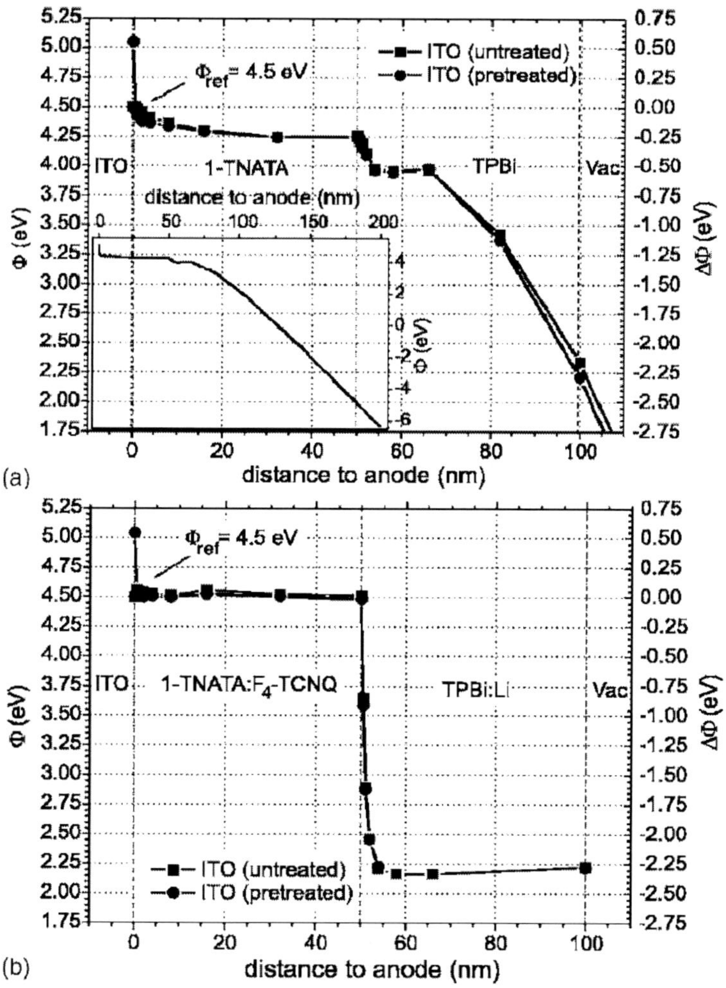

Reproduced from reference [37].

Figure 5. Work function obtained from KP measurements as a function of the film thickness, relative to the ITO substrate's work function for a) an undoped p-n heterojunction (inset: progression of the work function for a thicker TPBi film) and b) a doped p-n heterojunction.

Similar to other CGLs, the organic heterojunction CGLs also need effective vacuum level bending at the organic heterojunction CGL/ETL interface to insure good electron injection property. In 2008, Liao et al. reported tandem devices achieving both high efficiency and long lifetime using a strong electron acceptor material 1,4,5,8,9,11-hexaazatriphenylene hexacarbonitrile (HAT-CN) as interconnector [14, 15]. They employed an n-doped ETL (Li: BPhen) which can result in a large vacuum level bending at the ETL/HAT-CN interface to solve the electron injection problem. The tandem devices not only showed a double current efficiency of the referencing single unit device, but also exhibited higher power efficiency [14]. Moreover, the HAT-CN interconnector based tandem device showed rather high operating stability, almost triple times the lifetime of a single unit device (18000 h VS 6000 h) [15]. The work shows great potential of organic heterojunction type CGLs.

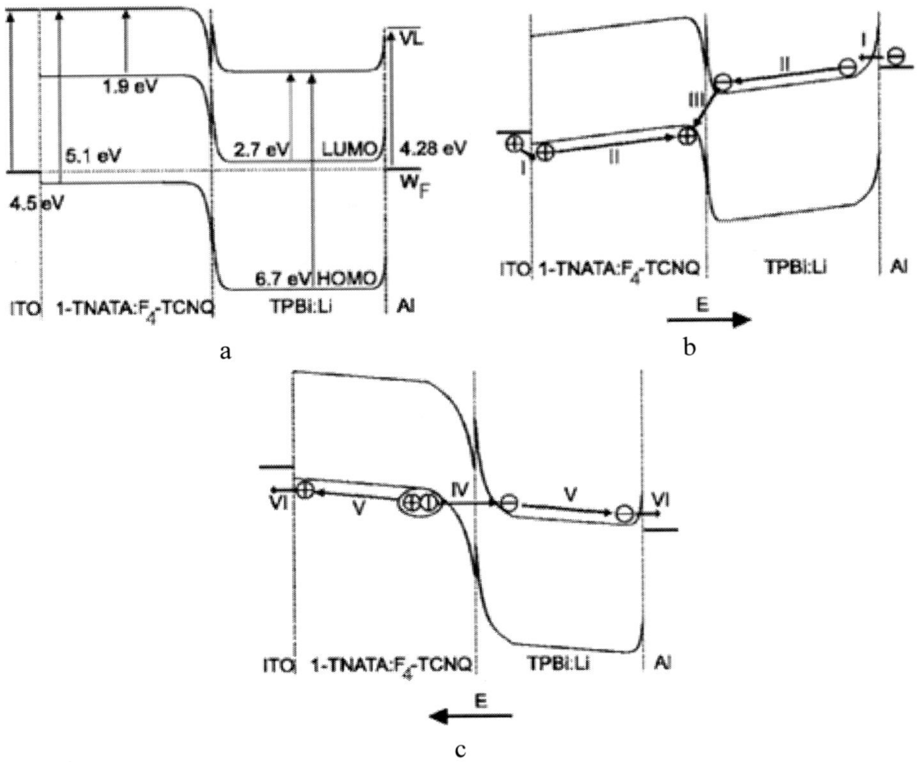

Reproduced from reference [37].

Figure 6. Energy-level alignment for a n-doped/p-doped CGL device: a) no external bias, b) under foreword bias (recombination), and c) under reverse bias (charge generation).

In recent years, it has become the new point of interests in the research on TOLEDs. High efficiency TOLEDs have been reported by using organic heterojunction as the CGLs, such as: C_{60}/pentacene, C_{60}/ZnPc, HAT-CN/NPB, HAT-CN/TAPC and HAT-CN/m-MTDATA [11, 14, 44-47]. When organic heterojunctions are employed as the CGLs, they usually contain two organic layers deposited in sequence to form donor-acceptor system, and the charges are generated at the interface under bias. The acceptor is usually a n-type organic material with big electron affinity, thus low LUMO. The choice of such organic semiconductors suitable for deposition is restricted. The commonly reported ones include F_4TCNQ, C_{60} and HAT-CN. The donor is usually a p-type material with good hole transport property, in most cases, the HOMO is around 5 eV. Because the dielectric constant of the organic semiconductor is usually low and the non-covalent electronic interactions between organic semiconductors are weak compared to inorganic semiconductors, two types of anisotype organic heterojunctions may be formed: accumulation and depletion organic heterojunctions. Considering the energy levels of acceptor and donor used in organic heterojunction CGLs, such as C_{60}/pentacene and HAT-CN/m-MTDATA, the Fermi level of the intrinsic p-type materials is higher than that of the n-type, so when they contact, the electrons would flow from p-side to n-side to form an accumulation type organic heterojunction, according to Yan et al. [48]. The formation of accumulation type organic heterojunctions is important for the performance improvement of resulting TOLEDs.

Different from the depletion junction in typical n-doped/p-doped CGL, where the depletion of free carriers will cause lower mobility, whereas the accumulation type organic heterojunctions will accumulate free holes at the p-side and electrons at the n-side, thus exhibit rather high mobility (several magnitudes higher than intrinsic layers) at both side [48, 49]. One significant advantage of such organic heterojunctions is that the accumulation of free carriers ensures the efficient charge generation, meanwhile, the bulk voltage drop across the organic heterojunction could be reduced to a minimum because of the high conductivity [49]. Due to these advantages, it is reasonable to expect better performances from the TOLEDs utilizing organic heterojunctions as the CGL. Actually, the most of high efficiency TOLEDs published so far are based on the organic heterojunctions as CGLs [12, 17, 46, 47, 50, 51].

4.1. Basics of Organic Heterojunctions

It is well known that semiconductors are defined by their unique electric conductivity behaviors, and can be classified into p-type and n-type semiconductors. A semiconductor heterojunction is the interface that occurs between p-type and n-type semiconductors, it is always advantageous to engineer the electronic energy bands in many solid state devices, including semiconductor light-emitting diodes, lasers, solar cells, and transistors. In fact, the concept of semiconductor heterojunctions had already been proposed and the energy-band profiles follow the Anderson model [52].

To date, all inorganic optoelectronic devices are based on this kind of semiconductor heterojunction. The most familiar heterojunction type in inorganic semiconductors is the depletion mode. In this type of heterojunction, a depletion junction is formed on either side of the heterojunction interface; namely, the positive charges are accumulated on the n-type side and the negative charges are accumulated on the p-type side in the depletion region. In this case, the internal electric field is opposite to the external field and the charges in the depletion region are immovable. However, the case of a heterojunction consisting of two organic semiconductors is somewhat different.

The organic semiconductor heterojunction (OHJ) [48] (i.e., n-type and p-type organic semiconductors) is a promising electronic system for charge recombination in OLEDs, for charge separation in organic photovoltaic cells (OPVs), and for charge accumulation in organic field-effect transistors (OFETs) owing to the energy mismatch between the frontier orbitals of the two organic semiconductors. Because the dielectric constant of the organic semiconductor is usually low and the non-covalent electronic interactions between organic semiconductors are weak compared to inorganic semiconductors, two types of anisotype heterojunctions may be formed: accumulation and depletion heterojunctions. The depletion heterojunction in organic semiconductors is similar to that in inorganic semiconductors. However, the accumulation heterojunction is completely different. In this case, the positive and negative charges are accumulated on the p-type and n-type sides, respectively, of the organic semiconductor to form the space charge region. We call this phenomenon a heterojunction effect, which has been well demonstrated in OFETs and OPVs exhibiting highly efficient device performance. The direction of the built-in voltage is from the p-type region to the n-type region.

More importantly, the accumulated charges within the charge region are movable. The accumulation of high density free charge carriers results in a high conductivity along the junction direction. To be able to effectively achieve and manipulate these processes in organic heterojunctions, the C_{60}/pentacene organic heterojunction is selected as an example to elucidate these processes. On the basis of the theory of thermal emission of electrons, the electron transfer from pentacene to C_{60} can be achieved since pentacene has a higher Fermi level than C_{60} in the flat band state (Figure 7 left). Also noted is that this charge transfer in turn contributes to the interfacial energy level equilibrium. Benefiting from the charge redistribution, the electrons and holes can be accumulated on n-type C_{60} and p-type pentacene, respectively, in the vicinity of the C_{60}/pentacene interface (Figure 7 middle).

Therefore, high-density free electrons and holes are provided at the C_{60}/pentacene junction (Figure 7 right); i.e., charge generation occurs. These generated charge carriers can move away from the interface in opposite directions under an external electric field. This process is beneficial to reducing the voltage drop across the CGL and hence the reduction of the overall driving voltage during the device operation. Obviously, the relative energy level of both semiconductor components is very important for the CGL construction, which directly determines the charge generation.

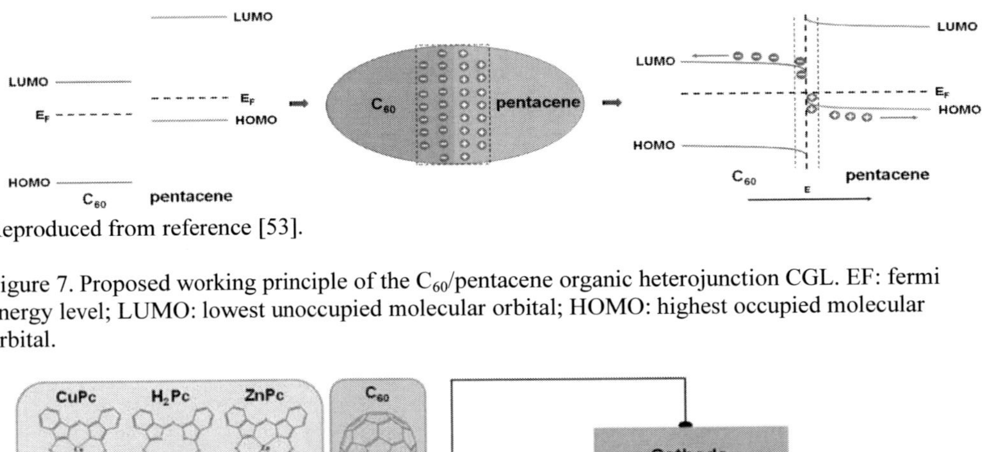

Reproduced from reference [53].

Figure 7. Proposed working principle of the C_{60}/pentacene organic heterojunction CGL. EF: fermi energy level; LUMO: lowest unoccupied molecular orbital; HOMO: highest occupied molecular orbital.

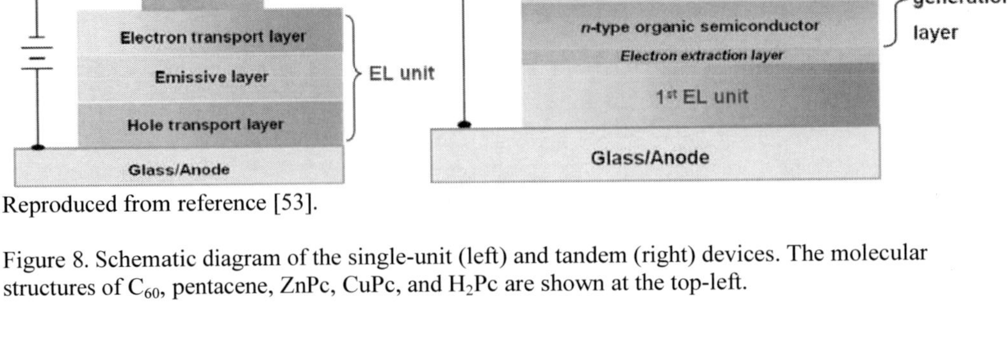

Reproduced from reference [53].

Figure 8. Schematic diagram of the single-unit (left) and tandem (right) devices. The molecular structures of C_{60}, pentacene, ZnPc, CuPc, and H_2Pc are shown at the top-left.

4.2. Bilayer Organic Heterojunction CGLs

Bilayer organic heterojunctions as CGLs have been used to fabricate high efficiency tandem OLEDs [53]. Figure 8 depicts the schematic diagrams of the corresponding single-unit and tandem OLEDs. The hole and electron transport layers are NPB and Alq_3, respectively. The emissive layer is Alq_3:C545T. In the bilayer organic heterojunction CGLs, the *n*-type layer is C_{60}, and the *p*-type layer is either pentacene, ZnPc, CuPc, or H_2Pc. The electron and hole extraction layers are LiF and MoO_3, respectively.

Figure 9 compares the EL characteristics of the C_{60}/pentacene organic heterojunction-based tandem OLED and the single-unit control OLED. It can be seen from Figure 9(a) that this tandem device exhibits a lower operational voltage, and the operational voltage ratio of this tandem device to the single-unit device gradually decreases with increasing current density and luminance. This demonstrates that the C_{60}/pentacene organic heterojunction exhibits remarkable characteristics as a connecting unit in tandem OLEDs.

As a result, not only is the current efficiency greatly enhanced from 15.2 cd/A for the single-unit device to 38 cd/A for the tandem device (Figure 9(b)), but the power efficiency of the tandem device is significantly improved, e.g., $tandem_{max}$: 21.9 lm/W *versus* single-$unit_{max}$: 10.1 lm/W (Figure 9(c)). This gives the best power efficiency improvement reported so far for tandem devices without any out-coupling technique, making the C_{60}/pentacene organic heterojunction CGL far superior to conventional CGLs.

Even at high luminance, the power efficiency enhancement is still significant; for example, under 1,000 cd/m^2, 10,000 cd/m^2, and 38,000 cd/m^2, the power efficiency enhancements are 1.74, 1.81, and 1.9 times that of the optimized single-unit, respectively. Actually, when the pentacene is replaced by H_2Pc, ZnPc, or CuPc, similar power efficiency improvements can also be achieved. Table 2 summarizes the detailed EL performance of all the devices here.

It can be seen that the organic heterojunctions as CGLs indeed exhibit excellent EL performance compared to the conventional ones.

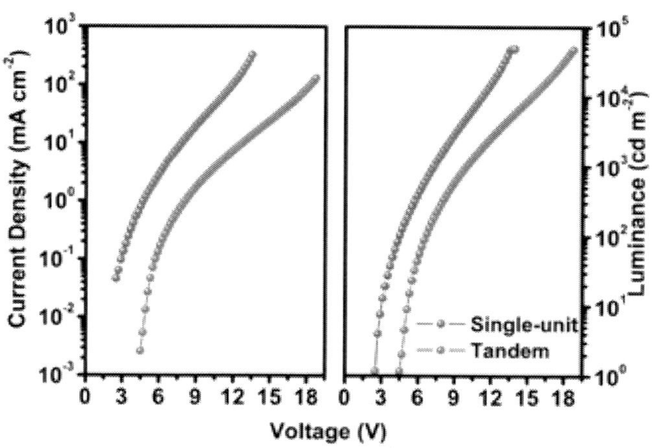

a

Figure 9. (Continued).

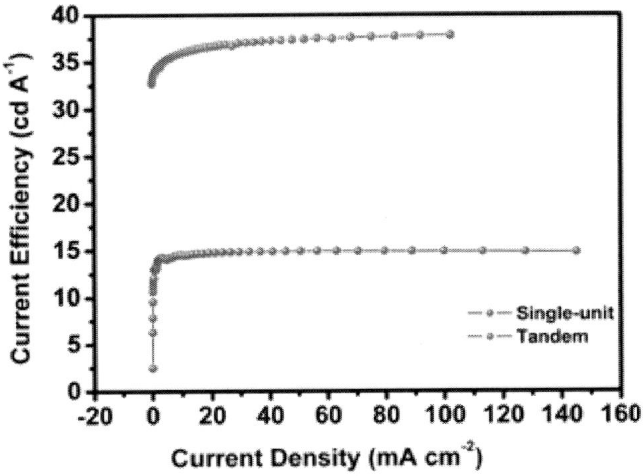

b

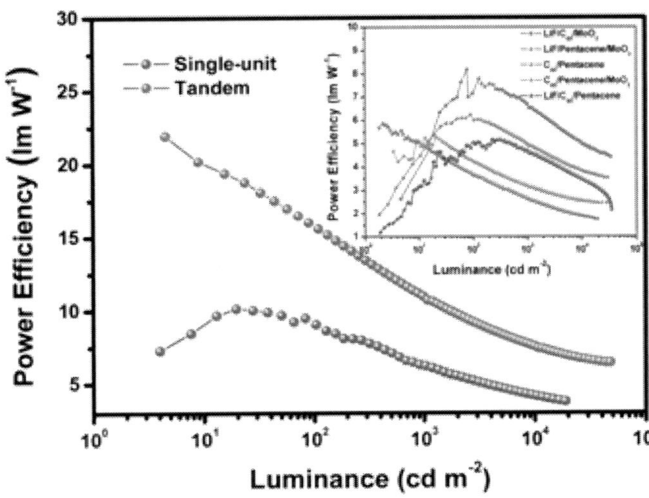

c

Reproduced from reference [53].

Figure 9. EL performance of the single-unit device and tandem device using a C_{60}/ pentacene OHJ CGL, including (a) The current density/luminance-voltage, (b) the current efficiency-current density, and (c) the power efficiency-luminance characteristics. Inset: the power efficiencies of the control tandem devices, in which the CGLs are assembled with different material combinations.

It is noted that the notable power efficiency improvement in the tandem device based on organic heterojunction CGLs is the key difference from the previously reported CGLs, where it was suggested to be difficult to achieve improvement in the high power efficiency. The power efficiency improvement in the C_{60}/pentacene-based tandem devices can be clarified by a complete understanding of the charge generation, transport, and extraction/injection occurring in the CGL. The direct experimental evidence is provided by investigating the capacitance–voltage (C–V) characteristics of the designed devices. As shown experimentally [53], the interface modified C_{60}/pentacene heterojunction-based device exhibited a gradual increase of capacitance above 5 V, which has been attributed to the effective charge transfer from pentacene to C_{60} due to the exceptionally strong acceptor characteristic of C_{60} molecules and the close energy levels of C_{60} and pentacene [54].

This fact conclusively proves that an electric-field-induced charge generation process takes place at the C_{60}/pentacene interface [55]. Recently, interfacial electronic structures of MoO_3 and LiF-modified pentacene/C_{60}-based charge generation layer (CGL) have also been investigated with photoemission spectroscopy well [56].

Figure 10 is the total schematic energy level alignment diagram at MoO_3 and LiF modified pentacene/C_{60}-based CGL. It is found that a small energy offset at the pentacene/C_{60} heterojunction makes it easy to transfer electrons from pentacene to C_{60} even under a small applied bias, facilitating the occurrence of charge generation.

The band bending observed in both pentacene and C_{60} is beneficial to exciton-dissociation and charge transport in opposite directions. At the MoO_3/pentacene interface, the high work function (W_F) of MoO_3 brings the highest occupied molecular orbital (HOMO) onset up to the Fermi level (E_F) not only for pentacene but also for most hole transport layer (HTL) materials of the adjacent electroluminescent (EL) unit as this CGL is connected into TOLED.

Therefore, holes can be efficiently injected from pentacene into this EL unit. Similarly, at the C_{60}/LiF interface, the low W_F of the LiF buffer layer makes the lowest unoccupied molecular orbital (LUMO) to pin close to the E_F not only for C_{60} but also for most electron transport layer (ETL) materials of the other EL unit, which induces the electrons to inject easily from C_{60} into that EL unit by tunneling through the thin LiF film. The favorable energy level alignment can effectively enhance charge generation, transport, and injection. The advantage of the MoO_3/pentacene/C_{60}/LiF structure is that thus formed CGL can greatly reduce the voltage drop and thus enhance the power efficiency (PE) of the corresponding TOLED.

It has been proven that HAT-CN as n-type semiconductor can also form effective organic heterojunction with some hole transport organic semiconductors such as m-MTDATA, TAPC and NPB, and high efficient TOLEDs can be achieved by using these organic heterojunction as CGLs [15, 34, 47, 57].

Table 2. Summary of device characteristics by exploiting single-unit and tandem architectures

CGL	V_t [V]	L [cd m^{-2}]	$\eta_{CE, max}$ [cd A^{-1}]	η_{CE} [cd A^{-1}]	$\eta_{PE, max}$ [lm W^{-1}]	η_{PE} [lm W^{-1}]
no	2.5	3033	14.9	14.8	10.1	5.1
C_{60}/pentacene	4.9	7322	37.8	36.6	21.9	7.9
C_{60}/ZnPc	4.9	7022	36.1	35.8	21	6.8
C_{60}/CuPc	5.1	6547	33.4	33.4	18.6	6.7
C_{60}/H_2Pc	5.3	6527	33.5	33.7	16	5.8
Li:BCP/MoO_3	5.5	6131	31.7	30.7	11.6	4.7
Li:BCP/MoO_3:NPB	5.5	5998	31	30.4	11.3	4.7

Reproduced from reference [53].

V_t: turn-on voltage examined at the luminance of 1 cd/m^2. L: luminance, η_{CE}: current efficiency, and η_{PE}: power efficiency obtained at the current density of 20 mA/cm^2. $\eta_{CE,max}$: maximum value of current efficiency. $\eta_{PE,max}$: maximum value of power efficiency. No: single-unit device.

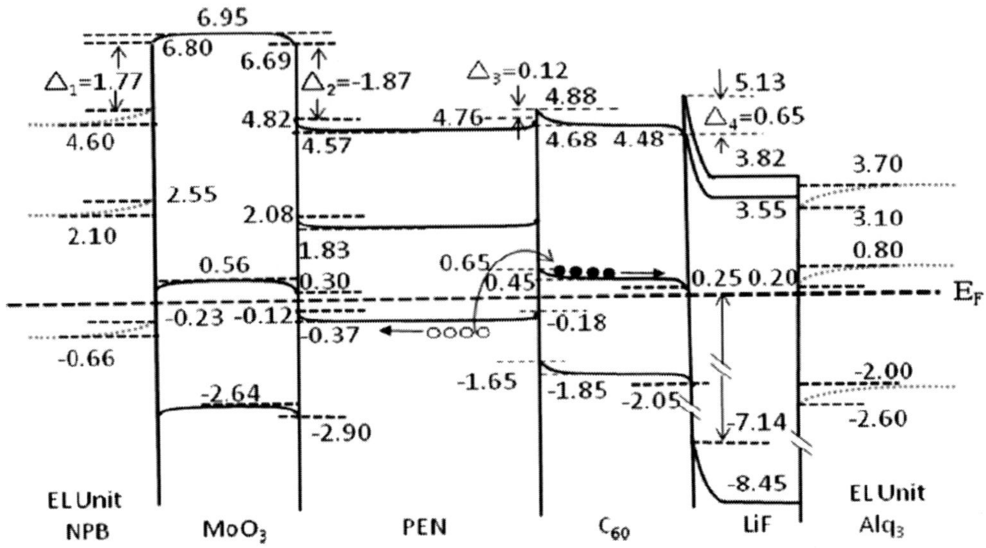

Reproduced from reference [56].

Figure 10. Schematic energy level alignment at the NPB/MoO_3/PEN/C_{60}/LiF/Alq_3 interfaces and proposed mechanisms of holes and electrons generation, where the dotted lines denote the energy levels of HTL and ETL in two adjacent EL units, and the solid dots denote electrons and the open dots denote holes. The data for VL, LUMO and HOMO denote the values of energy level with respect to the E_F.

It is well-known that HAT-CN is a strong electron accepter with a LUMO as low as 4.8 eV [48]. Considering that most commonly used hole transporting materials' HOMO are lying between 5~6 eV, HAT-CN is suitable to form a donor-acceptor system with them. In order to select an appropriate hole transporting material (HTM) to form organic heterojunction with HAT-CN, a set of devices have been prepared: ITO/MoO_3 (3 nm)/20 wt% MoO_3: TAPC (50 nm)/HTM (15 nm)/HAT-CN (15 nm)/3 wt% Cs: BPhen (50 nm)/Cs (1 nm)/Al (100 nm), here HTM includes m-MTDTA, TAPC, NPB and TCTA. Figure 11 gives a comparison in J-V properties of these devices [47]. The cases of NPB/HAT-CN and TAPC/HAT-CN have already been reported by other groups [15, 46].

It is well-known that the symmetry of J-Vcharacteristics is an indicator as the efficiency of charge generation at theinterface, i.e., the more symmetrical the J-V curve is, the better the charge generation will be [37]. As seen from Figure 11, in all the tested materials, HAT-CN/ m-MTDATA organic heterojunction shows the best symmetry as well as the largest current density at the same bias, indicating that m-MTDATA/HAT-CN organic heterojunction possess the best charge generation efficiency than the others. In addition, considering the wide optical band gaps of both materials (all more than 3 eV), the excess light absorption at visible spectrum range in this organic heterojunction could also be avoided when employing it as connnector in TOLED.

Based on HAT-CN/m-MTDATA organic heterojunction as CGL, high efficiency green TOLED with phosphorescent EL unit using $Ir(ppy)_2(acac)$ were achieved.

Its maximum power efficiency could reach 120 lm/W, yet remain 110.3 lm/W at 1000 cd/m^2, which is greatly enhanced from the corresponding single-unit device, shown in Table 3.

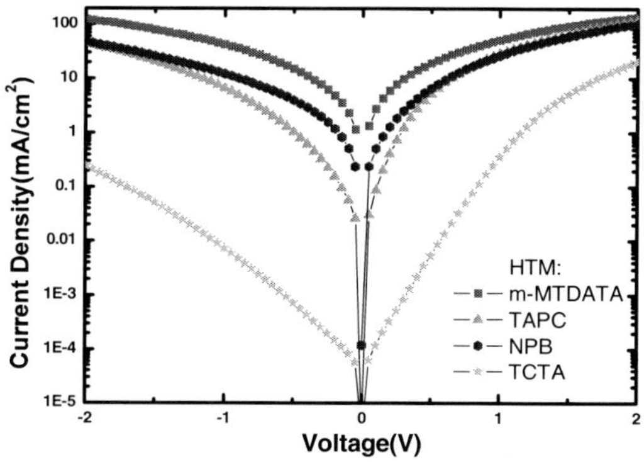

Reproduced from reference [47].

Figure 11. Comparison of J-V peroperties of given organic heterojunctions in the device structures of ITO/MoO_3 (3 nm)/20wt%MoO_3:TAPC (50nm)/HTM (15 nm)/HAT-CN (15 nm)/3 wt%Cs:BPhen (50 nm)/C_S (1 nm)/Al (100 nm). HTM is m-MTDATA, TAPC, NPB and TCTA, respectively.

Table 3. Comparison of single unit device and tandem device using HAT-CN/m-MTDATA CGL

Device	Von (V)	Voltage (V)	CE (cd/A)	PE (lm/W)	EQE (%)
Tandem	4.9	5.3/5.6/6.0	200.8/195.2/185.7	119.4/110.3/99.7	54.5/51.2/46.4
Single unit	2.4	2.6/2.8/3.2	80.4/81.8/72.7	96.8/92.8/70.8	21.9/22.2/19.5

Reproduced from reference [47].

The table shows the turn on voltage (V_{on}) and voltage (V), current efficiency (CE), power efficiency (PE), external quantum efficiency (EQE) at 100/1000/1000 (cd/m^2), respectively.

Also its maximum current efficiency and external quantum efficiency can reach as high as 201 cd/A and 54.5%, respectively, and only weakly drop to 195.2 cd/A and 51.2% at 1000 cd/ m^2, respectively, the best results published by now for TOLEDs. When two identical EL units as the reference single-unit OLEDs are connected by an CGL, the power efficiency of the resulting TOLEDs will stay unchanged due to the double operational voltage, even though the resulting current efficiency could be twice of the reference OLEDs.

Furthermore, if the CGL is not well designed, then the actual power efficiency will be even lower than that of the single unit OLEDs due to the extra voltage drop in the CGL. Therefore, in order to enhance the power efficiency of TOLEDs, one has to either increase the brightness or reduce the operational voltage at a certain current density. Actually, there are several ways to increase the current efficiencies in TOLEDs. A better current efficiency could be understood from two aspects: better exciton utilization and larger external quantum efficiency. First, for a well-designed TOLED, the charges are readjusted due to the insertion of interconnector, thus sometimes a better charge balance in tandem devices could be achieved than the referencing single unit devices. In addition, TOLEDs are much thicker than the single unit ones, naturally the leakage current would be reduced, and a higher current efficiency should be expected. Second, as reported, the fraction of radiation coupled into surface-plasmon modes would be less when the emitter is further away from the cathode, the maxima foreword luminance occurs roughly in the corresponding antinodes of the metal electrodes due to the interaction of microcavity and surface-plasmon modes [59-62].

So if we put the two emitting layers in TOLEDs roughly at the first and second antinode away from the aluminum cathode, then fine tuning the thickness of the charge transporting layers makes sure the emitting layers located at the best place to realize the maximum brightness, thus the external quantum efficiency can be enhanced. The above shows the possibility of TOLEDs achieving more than twice the current efficiency of the corresponding single unit devices.

The reduction of operational voltage in tandem devices is somehow a relative concept to compare the voltage drop in different CGLs. Because the injection of charges in an optimized single unit OLED is considered to be ideal ohmic, but the CGLs can't work better than that. This means that the extra voltage loss across the CGL should exist.

Obviously, the power efficiency in tandem devices is not possible to enhance with respect to the optimized single-unit devices. However, the TOLEDs achieving better power efficiencies than the referenced single unit ones have been reported by several groups using different CGLs [14, 15, 44, 45, 47, 49] and the current efficiencies are also improved by more than twice compared to the single unit ones.

In order to evaluate the energy loss in CGLs, we can define the ratio of energy conversion rate R as the achieved power efficiency (P_E) divided by the theoretical power efficiency (P_T) at a certain EQE for a given spectra:

$$R = \frac{P_E}{P_T} \tag{1}$$

If 1 watt electrical power is completely converted to 1 watt luminous power, then we can get:

$$P_T = K_m \times \frac{\int_{380nm}^{780nm} L(\lambda) V_\lambda d\lambda}{\int_{380nm}^{780nm} L(\lambda) d\lambda} \times \frac{EQE}{n} \tag{2}$$

where K_m = 683 lm/W, V_λ is the luminous efficiency function which reflects the spectral response of the human eye, $L(\lambda)$ is the EL spectra intensity of the OLED and n is the number of EL units in the TOLED (for single unit, n = 1).

The advantage of using R to evaluate the performance of CGLs is that it can exclude the effects of EQE related factors, such as charge balance, spectra shifting and out-coupling rate, because R is only related to the percentage of electrical power converted to light emission in an OLED. Taken an example [11], two white TOLEDs with different CGLs of TPBi:Li_2CO_3/ TCTA:MoO_3 and C_{60}/pentacene were fabricated. We can calculate that the former TOLED has R = 63%, and the later TOLED has R = 71%, meaning that the C_{60}/pentacene CGL indeed has less energy loss than the n-doped/p-doped one, considering the energy consumption in other parts of the two TOLEDs is the same.

Theoretically, the R values of the single unit OLEDs and the tandem OLEDs with two same units as the corresponding single unit should be approximately same only when the CGL in tandem devices can efficiently generate charges. As the TOLEDs based on HAT-CN/ m-MTDATA as the CGL, its R = 83.2%, which is very close to the R = 83.5% of single unit OLEDs. This means that HAT-CN/m-MTDATA organic heterojunction is indeed a very optimum CGL in charge generation and charge injection.

4.3. Bulk Organic Heterojunction CGLs

It is found that the charge generation originating from the charge transfer from p-type organic semiconductor to n-type organic semiconductor is a prerequisite and is very important to the organic heterojunction CGL operation. For the bilayer organic heterojunction CGL systems, the charge generation occurs at the narrow interface between two organic semiconductors, which may limit the generation of charge carriers and also cause the accumulation of the generated charge carriers.

As we know, beside the bilayer heterojunctions, the bulk heterojunctions have widely been used in the fabrication of high efficiency organic solar cells (OSCs) [63-65]. The bulk organic heterojunction by blending the donor and acceptor organic materials was first proposed by Heeger [63], and have been widely used to achieve high-efficiency OSCs where the blend of a p-type organic semiconductor and an n-type organic semiconductor shows phase states of only several nanometers in microcosmic sizes, leading to an easily effective carrier diffusion. Up until now, almost all high-performance OSCs were fabricated by using bulk organic heterojunctions, such as PCDTBT:$PC_{70}BM$ [64], PTB_7:$PC_{71}BM$ [65] and ZnPc: C_{60} [66]. The bipolar charge generation characteristics of the bulk organic heterojunctions imply that the bulk organic heterojunctions could also be used as excellent CGLs for the fabrication of high performance tandem OLEDs.

As shown, the bulk heterojunctions can indeed be used as CGLs to fabricate high efficiency TOLEDs. Figure 12 presents the schematic diagram of the cross sectional view of the fabricated TOLED based on ZnPc:C_{60} bulk organic heterojunction as CGL [17, 45].

One key prerequisite for designing the ZnPc:C_{60} bulk organic heterojunction CGL is the effective control of ratios between ZnPc and C_{60}. Considering the film morphology and transport characteristics, the ZnPc:C_{60} blend composition is fixed at a weight ratio of 1:1 by optimization. Thus the structure of the fabricated TOLEDs based on a ZnPc:C_{60} bulk organic heterojunction CGL is ITO/MoO_3 (6 nm)/NPB (90 nm)/Alq_3:C545T (30 nm)/Alq_3 (30 nm)/ LiF (0.3 nm)/ZnPc:C_{60} (1:1, 30 nm)/MoO_3 (3 nm)/NPB (50 nm)/Alq_3:C545T (30 nm)/Alq_3 (30 nm)/LiF (1 nm)/Al (120 nm). As shown in Table 4, the maximum current efficiency of the tandem OLEDs using a C_{60}:ZnPc bulk organic heterojunction CGL reaches 36.5 cd/A, which is 2.4 times higher than that of the single-unit device (15.2 cd/A). In addition to the improvement in current efficiency, the bulk bulk organic heterojunction CGL also significantly improves the power efficiency of the resulting tandem OLED to 21 lm/W, which is 2 times that of the single-unit device (10.1 lm/W). These results are similar to those of the bilayer OHJ CGL-based OLEDs. This means that the ZnPc:C_{60} bulk organic heterojunction can indeed also function as an effective bipolar CGL. Actually, the ZnPc:C_{60} bulk organic heterojunction CGL is an incremental step from the bilayer organic heterojunction CGL towards generalization of the concept. However, for the case of the ZnPc:C_{60} bulk heterojunction CGL, the ZnPc and C_{60} molecules can sufficiently contact each other, and thus the charge transfer occurs in the whole ZnPc:C_{60} bulk heterojunction, which is completely different from the bilyaer organic heterojunction CGLs where the charge transfer occurs at the interface between two organic semiconductors. This indicates that the charge transfer in bulk heterojunction CGLs should occur more easily than in bilayer organic heterojunction CGLs, thus generating more charges. This is proven by the achievement of higher current and power efficiencies in TOLEDs based on ZnPc:C_{60} bulk organic heterojunction CGL versus in tandem devices with ZnPc/C_{60} bilayer organic heterojunction CGL.

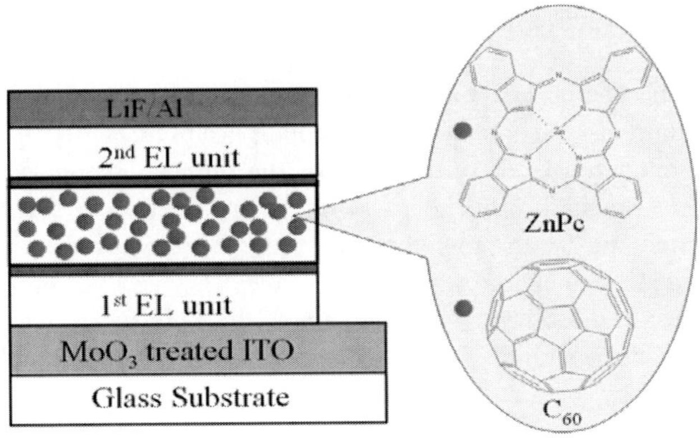

Reproduced from reference [45].

Figure 12. Schematic diagram of the cross sectional view of a tandem OLED made with a ZnPc:C_{60} OHJ CGL, together with the molecular structures of ZnPc and C_{60}. The ultra-thin interlayers of LiF (blue) and MoO_3 (red) are introduced to achieve an efficient electron and hole injection, respectively.

Table 4. Summary of the EL performance of all tandem devices using a bulk organic heterojunction as CGL

CGL	V_t [V]	L [cd/m^2]	$\eta_{CE, max}$ [cd/A]	η_{CE} [cd/A]	$\eta_{PE, max}$ [lm/W]	η_{PE} [lm/W]
C_{60}:ZnPc	4.9	7020	36.1	36	21	6.8
C_{60}:CuPc	5.1	6445	33.4	31.5	22.4	5.6
C_{60}:H_2Pc	5.5	6623	34.8	34.4	16	5.8

Reproduced from reference [17].

V_t: turn-on voltage examined at the brightness of 1 cd/m^2. L: luminance, η_{CE}: current efficiency, and η_{PE}: power efficiency obtained at the current density of 20 mA/ cm^2. $\eta_{CE, max}$: maximum value of current efficiency. $\eta_{PE, max}$: maximum value of power efficiency.

The charge carrier mobility then begins to play a major role in the bulk organic heterojunctions. As we know, the mobility of electrons and holes in bulk organic heterojunctions will be lower than those in the pure materials used in bilayer organic heterojunction [67]. If the mobility is not sufficiently high, the charge carriers will not be transported rapidly by the external electric field.

As a result, the excess charges will recombine again or remain in the device as undesirable space charges that oppose the drift of new charge carriers. The latter problem can occur if the electron and hole mobilities are highly imbalanced such that one species is much more mobile than the other. In this case, the device performance will be worse. In fact, it is found that the mobility of the organic semiconductors used in the organic heterojunction CGLs strongly influences the device performance [49].

The other bulk organic heterojunctions as CGLs (e.g., CuPc:C_{60} and H_2Pc:C_{60}) are also used to further demonstrate the effectiveness of the design concept of bulk organic heterojunctions as CGLs. It is found that a similar power efficiency improvement can also be achieved (Table 4), indicating that the bulk organic heterojunctions as CGLs are also a viable option for the fabrication of high-performance TOLEDs.

Because the mobility in bulk heterojunctions is lower than that in bilayer organic heterojunctions, the structure consisting of a bulk organic heterojunction sandwiched between homogeneous layers of p-type organic semiconductor and n-type organic semiconductor is also explored as CGL in TOLEDs [17, 57]. An advanced example that HAT-CN/HAT-CN: TAPC/TAPC is used as CGL is obtained, where a high efficiency green phosphorescent TOLED is successfully fabricated [57]. It can be seen that the utilization of the HAT-CN: TAPC blending layer between HAT-CN and TAPC in a CGL greatly enhances charge generation and separation, which was found by carrying out current–voltage analyses. This means that the three-layer structure is more suitable as a connecting unit for the fabrication of high efficiency TOLEDs. Clearly, the TOLEDs based on HAT-CN/HAT-CN:TAPC/TAPC as CGL showed more than double enhancement in current efficiency with respect to the corresponding single-unit device, which is also higher than that of TOLEDs with HAT-CN/ TAPC as CGL. More encouragingly, the power efficiency of TOLEDs with this sandwiched CGL was enhanced by about 1.3–1.5 times, reaching 113 lm/W at 1000 cd/m^2 brightness as compared to 76 lm/W from the single-unit OLEDs and 86 lm/W from the HAT-CN/TAPC-based TOLEDs. The TOLEDs with the new CGL also showed good stability, at least enhancing 4.5 times compared with the single unit devices.

4.4. Effects of Interfacial Energy Level and Charge Carrier Mobility

As shown above, the design concept of organic heterojunctions as CGLs is strikingly different from that of the conventional CGLs. The unique cooperation of charge generation, transport, and extraction/injection processes occurring in the organic heterojunctions remarkably reduces the operational voltage, rendering an effective method for high power efficiency TOLEDs. As demonstrated, it is necessary to first form an accumulation-type space charge region at the organic heterojunction interface where the holes are accumulated on the *p*-side and the electrons are accumulated on the *n*-side, and then the separated charge carriers can effectively be transported and injected into the corresponding EL units, which are directly related to the energy levels and the mobility of charge carriers in the organic heterojunctions [68]. To verify this, a homologous series of *p*-type thiophene organic semiconductors NaTn (naphthyl end-capped oligothiophenes, n = 2-6 represents the number of thiophene units are used, see the inset of Figure 13b for the molecular structures) [69, 70] as the p-type organic semiconductor in the organic heterojunction CGL to study these effects. The C_{60} was employed as the *n*-type organic semiconductor (C_{60}/NaTn) in the devices. The NaTn organic semiconductors possess significant differences in HOMO levels (Figure 13a) and hole mobilities (Figure 13b), which make them suitable for investigating our hypothesis. Five green TOLEDs with NPB/Alq_3:C545T/Alq_3 active layers based on C_{60}/NaTn OHJ CGLs were fabricated. Table 5 summaries the detailed EL performance of the five fabricated TOLEDs. It can be seen that the current efficiency of the tandem devices is over 2 times that of the single-unit device at a given current density, demonstrating the C_{60}/NaTn organic heterojunctions functioned as an effective bipolar CGL. More importantly, the power efficiency of the tandem devices is also remarkably enhanced. The maximum power efficiencies from C_{60}/NaT2- to C_{60}/NaT6-based tandem devices are 17.6 lm/W, 18.2 lm/W, 21.5 lm/W, 18.4 lm/W, and 20.5 lm/W, respectively, which are nearly 1.7-2 times that of the single-unit OLED (10.1 lm/W).

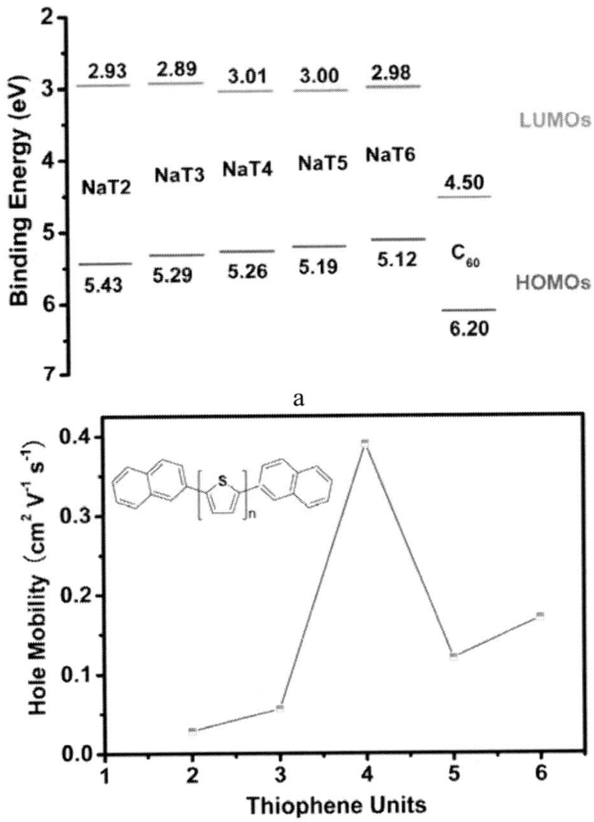

The HOMO levels and hole mobilities are obtained from reference [70]. Reproduced from reference [68].

Figure 13. (a) Energy levels of NaTn and C_{60}. (b) Hole mobility of NaTn. Inset: molecular structure of NaTn, n = 2-6.

This confirms again that the design concept of organic heterojunctions as CGLs is a predominant technique for achieving high power efficiency in TOLEDs. However, the five tandem devices show a marked difference in luminance and efficiency at the same current density. As shown, the NaT4 leads to the highest efficiency, followed by NaT6, NaT5, NaT3, and NaT2. This means that the energy levels and the mobility of the C_{60}/NaTn organic heterojunction have strong effects on the performance of the fabricated TOLEDs, which are directly related to the processes of charge generation and charge transport. Since the generated charge carriers stem from the electron transfer from the Fermi level of the NaTn to the Fermi level of the C_{60}, the position of the HOMOs for NaTn (ranging from 5.43 eV for NaT2 to 5.12 eV for NaT6 (Figure 12a)) relative to the LUMO of C_{60} probably contributes to the different capabilities of charge generation in the C_{60}/NaTn CGL. To further elucidate this problem, the capacitor devices based on C_{60}/NaTn were fabricated and their

C-V characteristics were measured, which are shown in Figure 14a. It is clearly seen that the capacitance is gradually decreased with increasing energy-level difference between the HOMOs of NaTn and the LUMO of C_{60} in the saturation regime, as shown in Figure 14b.

Because the magnitude of the capacitance can qualitatively give the charge generation capability, the variation regulation of the capacitance with the energy levels between the two organic semiconductors used means that the HOMO level of *p*-type organic molecules close to the LUMO of *n*-type organic molecules will generate more charge carriers in the CGL. Correspondingly, a high EL efficiency will then be obtained. As shown in Figure 14c, the maximum current efficiency and power efficiency does gradually increase from NaT2 to NaT6 by the energy level regulation, except for the case of NaT4.

Because the NaT4 possesses the highest hole mobility (0.39 cm^2/Vs, Figure 14d) with respect to the others, the achievement of the highest efficiency in TOLEDs with NaT4-based CGL indicates that the high mobility of the organic semiconductors used in organic heterojunction CGLs is even more important.

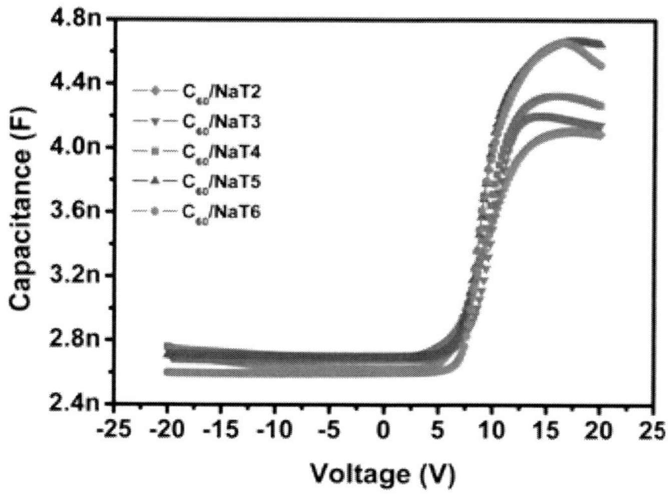

a

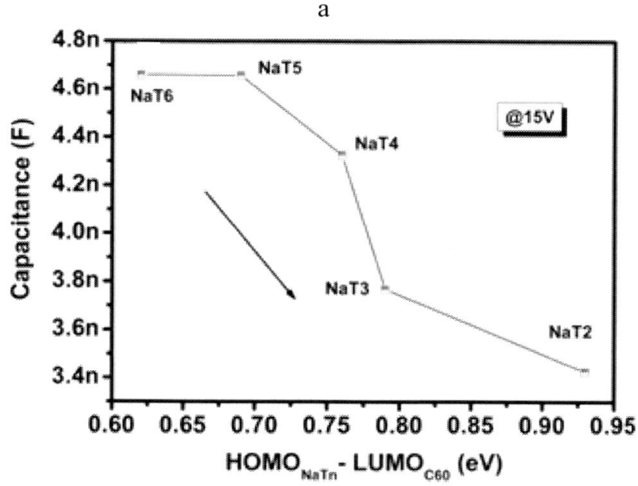

b

Figure 14. (Continued).

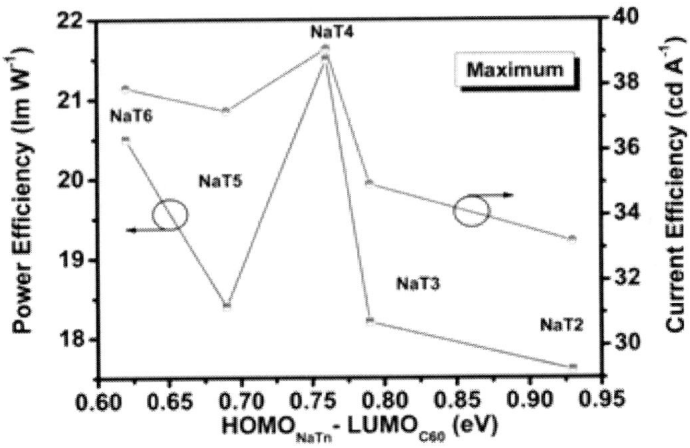

c

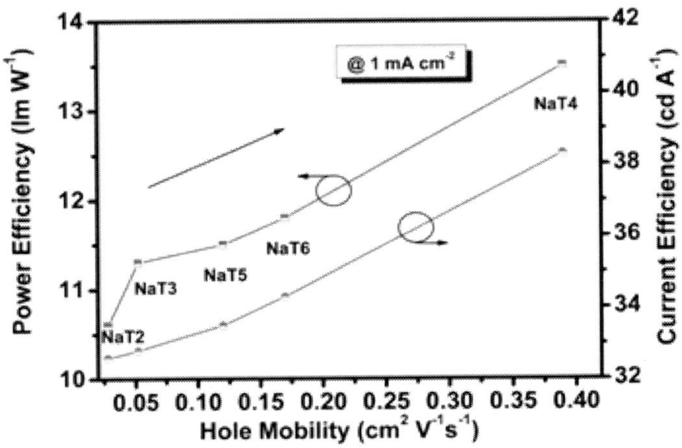

d

Reproduced from reference [68].

Figure 14. (a) *C-V* characteristics of devices based on C_{60}/NaTn OHJs at a fixed frequency of 1 kHz. The device structures are ITO/LiF (100 nm)/Alq_3 (30 nm)/LiF (0.3 nm)/C_{60} (20 nm)/NaTn (10 nm)/MoO_3 (3 nm)/NPB (50 nm)/LiF (100 nm)/Al (120 nm). (b) Capacitance versus the difference between the HOMOs of NaTn and the LUMO of C_{60} at 15V in the saturation regime. (c) Maximum power efficiency and current efficiency versus the difference between the HOMOs of NaTn and the LUMO of C_{60}. (d) Characteristics of power efficiency and current efficiency versus the hole mobility of NaTn at the current density of 1 mA cm^{-2}.

The high mobility will rapidly transport the generated charges towards the interfaces adjacent to the EL units under the electric field, thus facilitating the reduction of operational voltages. As shown in Figure 14d, the current efficiency and power efficiency of the fabricated TOLEDs based on NaTn/C_{60} organic heterojunction s as the CGL indeed increase with the hole mobility of NaTn from NaT2, NaT3, NaT5, NaT6, to NaT4.

Therefore, the organic semiconductors used to construct high-performance organic heterojunction CGLs not only need to have proper energy levels, i.e., closer levels between the HOMO of *p*-type organic semiconductor and the LUMO of *n*-type organic semiconductor, but also exhibit higher mobilities.

Table 5. Summary of the EL characteristics of the five fabricated tandem OLEDs based on C_{60}/NaTn OHJ CGLs

CGL	V_t [V]	L [cd m^{-2}]	$\eta_{cd, Max}$ [cd A^{-1}]	η_{cd} [cd A^{-1}]	$\eta_{p, Max}$ [lm W^{-1}]	η_p [lm W^{-1}]	EL_{peak} [nm]
C_{60}/NaT2	4.9	6620	33.2	33.1	17.6	6.4	524
C_{60}/NaT3	4.9	7000	35	34.8	18.2	6.7	524
C_{60}/NaT4	4.9	7820	39.1	39.1	21.5	7.8	524
C_{60}/NaT5	4.9	7300	37	36.5	18.4	6.9	524
C_{60}/NaT6	4.9	7440	37.9	37.2	20.5	7.3	524

Reproduced from reference [68].

Turn-on voltage V_t is tested at 1 cd m^{-2} (V_t); Luminance (L), current efficiency (η_{cd}), power efficiency (η_p), and the peak of EL spectra (EL_{peak}) are tested at 20 mA cm^{-2}; Maximum current efficiency ($\eta_{cd, Max}$); Maximum power efficiency ($\eta_{p, Max}$).

The results offer a design/selection rule for organic semiconductors used to construct effective organic heterojunction CGLs, which is useful to fabricate high performance TOLEDs in the future.

4.5. Mechanism of Charge Generation in Organic Heterojunctions

To elucidate the mechanism of charge generation in organic heterojunctions, using m-MTDATA/HAT-CN as an example, the following device ITO/MoO_3 (3 nm)/20 wt%MoO_3: TAPC (50 nm)/m-MTDATA (15 nm)/HAT-CN (15 nm)/3 wt%Cs:BPhen (50 nm)/Cs (1 nm)/ Al (100 nm) (device A) was fabricated and the I-V characteristics at different temperatures were measured.

The p-doping/organic heterojunction/n-doping structure is introduced to simulate the sequences in a real tandem device (when under reverse bias, the current flows just like the tandem device fabricated above), also to ensure ohmic contact at the electrodes [71]. Considering the high conductivity of MoO_3:TAPC and Cs:BPhen doping layers, the voltages would mainly drop on the m-MTDATA/HAT-CN heterojunction [72]. Thus the J-V characteristics of the device will be determined by the electrical properties of the heterojunction. When under reverse bias, the current is mainly generated in the organic heterojunction rather than injection from electrodes. Because of using the doped hole and electron transporting layers at both sides of the electrodes, the injection of opposite charge carriers should be very hard. The diffusion based Shockley model of heterojunctions predicts a linear relation in lnJ~V plot [73]:

$$J = I_0\left[\exp\left(\frac{eV}{nKT}\right) - 1\right] \quad (1)$$

which is not found in above device. However, as shown in Figure 15, this device gave a linear relation of lnJ~T rather than lnJ~1/T indicated by Shockley model (Figure 15 inset). Therefore, it is proposed that the charge generation process in m-MTDATA/HAT-CN organic heterojunction is charge tunneling, which can be described by an experimental tunneling equation for heterojunctions under reverse bias [74].

$$J_r = G_0 V_a \exp[-U(V_d + V_a)^{-1/2}] \quad (2)$$

J_r is the reverse current density, G_0 is a constant determined by the nature of the material, U is a linear variable of temperature, V_d is the built-in potential, V_a is the applied reverse voltage (written as a positive value).

The equation is initially developed from Zener tunneling equation to describe the electron tunneling from the valence band of a p-type semiconductor to the conduct band of an n-type semiconductor in staggered gap type heterojunctions [75-77]. A double logarithm J-V plot of Device A under various temperatures is given in Figure 16.

It can be clearly divided into two regions: under small voltages (~0.1 V), all curves have the same slope of approximately 1, meaning an ideal ohmic contact and transportation, which can be contributed to the leakage current under small bias; while under higher voltages, the curves show a power law relation and can be nicely fitted by (3) at various temperatures, from 137 K to 297 K. By far, it has become obvious that the J-V properties of the m-MTDATA/ HAT-CN organic heterojunction should be explained by the charge tunneling model rather than the Shockley equation. Accordingly, a schematic diagram explaining the working principles of HAT-CN/m-MTDATA organic heterojunction is given in Figure 17. The band bending is depicted according to its charge accumulating property at equilibrium.

When under certain bias, the electrons in the HOMO of m-MTDATA would tunnel into the LUMO of HAT-CN, thus generate an electron in the LUMO of HAT-CN and leaves a hole in the HOMO of m-MTDATA. Then the generated holes and electrons get transported and injected to the adjacent layers under the influence of electric field.

Consequently, the efficient charge generation and ohmic injection of HAT-CN/m-MTDATA organic heterojunction are guaranteed by its tunneling nature, further resulting in the high energy conversion rate in the TOLEDs.

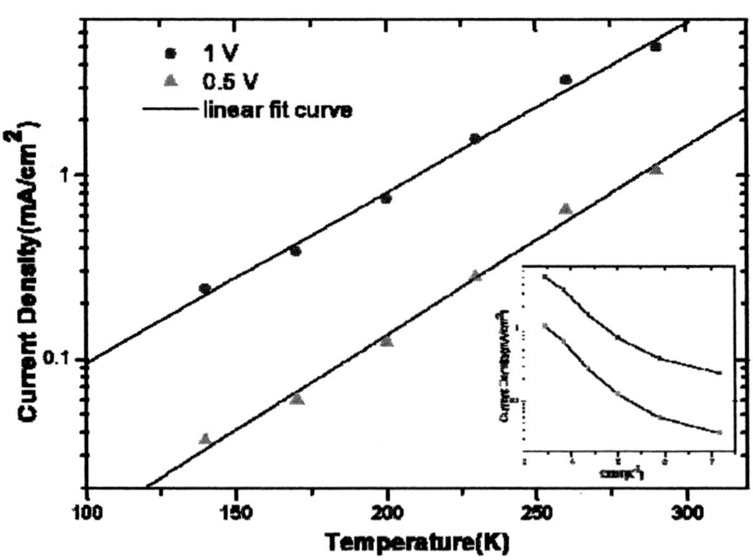

Reproduced from reference [47].

Figure 15. Low temperature dependence of current density at a fixed reverse bias of 0.5 V and 1 V. Inset: plot of J~1000/T.

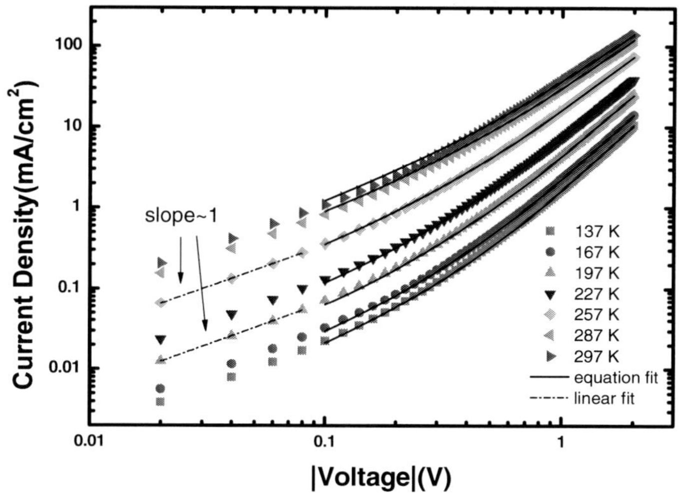

Reproduced from reference [47].

Figure 16. Fitting results of device A under various temperatures.

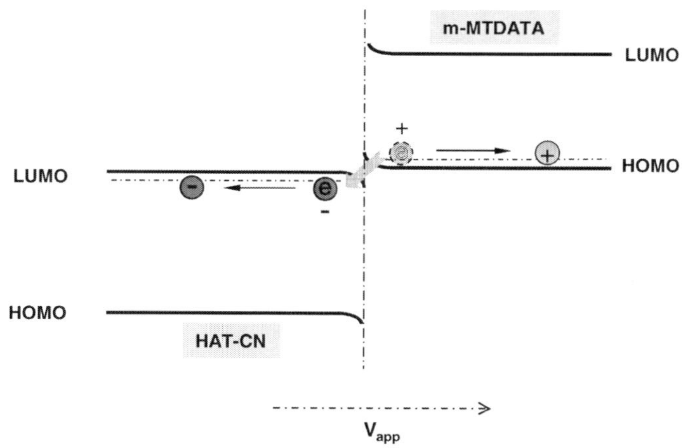

Reproduced from reference [47].

Figure 17. A schematic diagram for the charge generation process in m-MTDATA/HAT-CN organic heterojunction.

These results have fully revealed the advantages of organic heterojunctions as CGLs in high power efficiency TOLEDs. The commercial productions require OLEDs to achieve both high efficiency and long lifetime, there is no doubt that TOLEDs will continue to be the focus of OLEDs research, and keep progressing towards better performance.

4.6. Tandem White OLEDs Based on Organic Heterojunction CGLs

As we know, TOLEDs are technologically interesting because not only can the luminance and current efficiency be improved linearly with the number of electroluminescent (EL) units in the tandem OLED, but also leakage current and breakdown of the electric field can be avoided due to the higher luminance at a low current density and the thicker organic films, resulting in a long lifetime. Importantly, the state-of-the-art tandem OLEDs are very easy to vertically stack either individual red, green, and blue emission units or multiple white emission units in series via CGLs to achieve white emission [78].

As demonstrated above, using organic heterojunctions as CGLs can enhance the power efficiency of the fabricated TOLEDs. Accordingly, the tandem white OELDs based on organic heterojunctions as CGLs were fabricated [79, 80], and found that the power efficiency of the device was indeed enhanced and the efficiency roll-off was greatly reduced compared to the conventional CGL-based tandem white OLEDs.

Figure 18 shows the device structures of the tandem WOLEDs based on conventional TPBi:Li_2CO_3/TCTA:MoO_3 CGLs (Figure 18(a)) and C_{60}/pentacene organic heterojunction CGL (Figure 18(b)). The white emission is realized by stacking a green-red phosphorescence unit and a blue phosphorescence unit via CGLs.

Figure 19 shows the EL characteristics of both tandem white OLEDs based on TPBi: Li_2CO_3/TCTA:MoO_3 (tandem A) and C_{60}/pentacene OHJ (tandem B) as CGLs.

As shown in the inset of Figure 19(a), tandem A has the turn-on voltage of 5.5 V and the operational voltages of 6.7 V at 100 cd/m_2 and 8.3 V at 1000 cd/m^2. However, the turn-on voltage is reduced to 5.1 V and operational voltage of 6.0 V 100 cd/m^2 and 6.9 V at 1000 cd/m^2 as using C_{60}/pentacene organic heterojunction as the CGL (tandem B). This indicates that the C_{60}/pentacene organic heterojunction should have the further better electrical properties as the internal floating connecting electrode than TPBi:Li_2CO_3/TCTA:MoO_3. As a result, the maximum current efficiency, external quantum efficiency, and power efficiency reach 101.5 cd/A, 45.7%, and 53.8 lm/W, respectively (Figures 19(a) and (b)).

As seen, although the TPBi:Li_2CO_3/TCTA:MoO_3 as CGL indeed also leads to the high current efficiency and external quantum efficiency, which reach 98.6 cd/A and 44.9% (Figure 19(b)), respectively, indicating an effective CGL as previously reported, but the maximum power efficiency of 48.7 lm/W is low (Figure 19(a)) due to the high operational voltage.

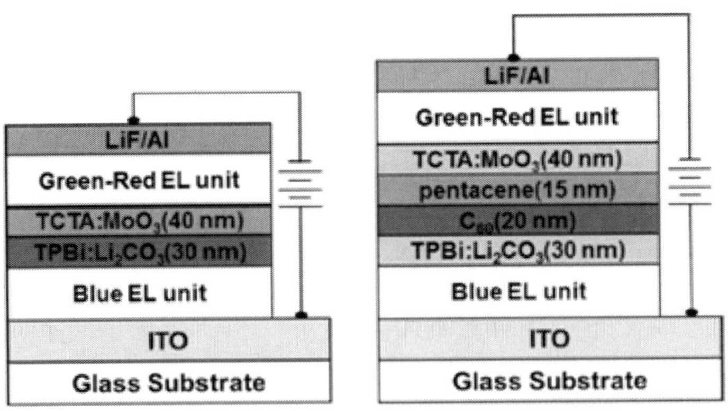

a b

Reproduced from reference [79].

Figure 18. Schematic diagrams of the tandem white OLEDs based on different CGLs. (a) TPBi: Li_2CO_3/TCTA:MoO_3 as the CGL. (b) C_{60}/pentacene organic heterojunction as the CGL.

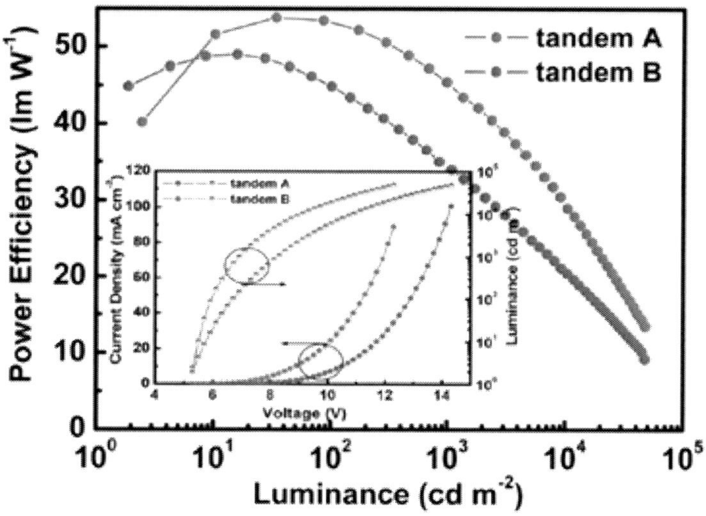

a

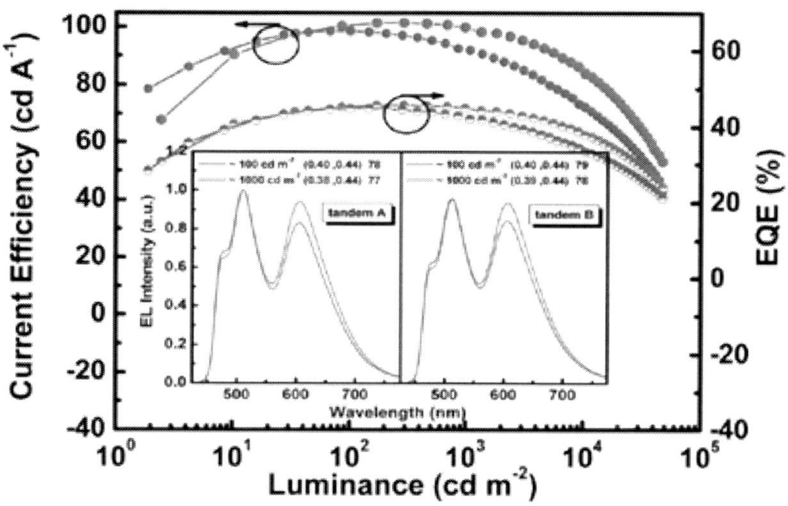

b

Reproduced from reference [79].

Figure 19. (a) Power efficiency versus luminance characteristics. Inset: J-V-L characteristics. (b) Current and external quantum efficiency versus luminance characteristics. Inset: EL spectra detected at the luminance of 100 cd/m^2 and 1000 cd/m^2.

More importantly, the utilization of C_{60}/pentacene organic heterojunction as the CGL significantly improves the efficiency roll-off of fabricated tandem white OLEDs. The efficiency is slightly reduced to 101 cd/A, 45.5%, and 53 lm/W and 99.9 cd/A, 45%, and 45 lm/W at 100 cd/m^2 and 1000 cd/m^2, respectively. The power efficiency roll-off values are only 1.5% at 100 cd/m^2 and 16% at 1000 cd/m^2.

Comparatively, the serious efficiency roll-off is given in TPBi:Li_2CO_3/TCTA:MoO_3-based tandem white OLEDs, where the power efficiency is decreased to 34 lm/W at 1000 cd/ m^2 with the roll-off value of 24%.

As previously demonstrated [53, 68], the effective charge transfer from pentacene to C_{60} will result in the accumulation of the holes on the p-side and the electrons on the n-side in the C_{60}/pentacene organic heterojunction CGL. The formation of the accumulation-type space charge region not only supplies the amounts of charges used to recombine, but also leads to the formation of a high conductance region, thus greatly reducing the voltage drop in connecting layer. The generation of amounts of charges in C_{60}/pentacene organic heterojunction CGL should also be favor to improving charge balance, thus efficiency roll-off. The inset in Figure 19(b) shows the EL spectra of the two tandem white OLEDs at 100 and 1000 cd/m^2. The similar EL spectra of tandem A and B devices indicate that inserting the C_{60}/pentacene organic heterojunction does not lead to additional optical interference and microcavity effect.

5. APPLICATION OF TANDEM WHITE OLEDS

Due to the advantages of high efficiency and long lifetime of tandem OLEDs, tandem white OLEDs for lighting applications have well been fabricated by some companies, such as LG, First O-Lite, Panasonic, Sharp, etc. The power efficiency of over 100 lm/W in tandem white OLEDs have successfully realized by combining light extraction technology [81-85]. LG Chem. has also commercialized different sizes OLED lighting panels with a power efficiency of 60 lm/W and a lifetime of 40000 hrs (Figure 20) [86]. Figure 21 gives the real applications of OLED lighting panels in protected eye lamps and library`s reading lights by LG Chem [86]. Figure 22 gives the schematic diagram of 3-tandem OLED lighting devices with internal and external light extraction layers fabricated by LG.

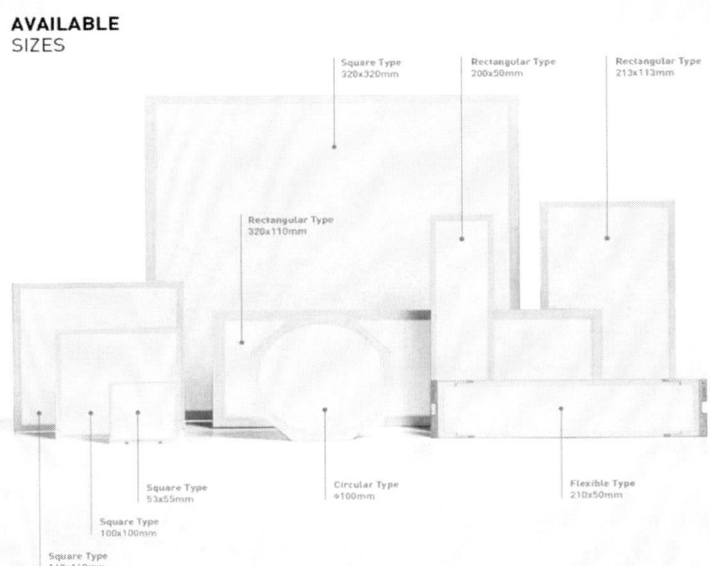

Reproduced from reference [86].

Figure 20. Different sizes OLED lighting panels realized by LG Chem.

Reproduced from reference [86].

Figure 21. OLED lamps and reading lights by LG Chem.

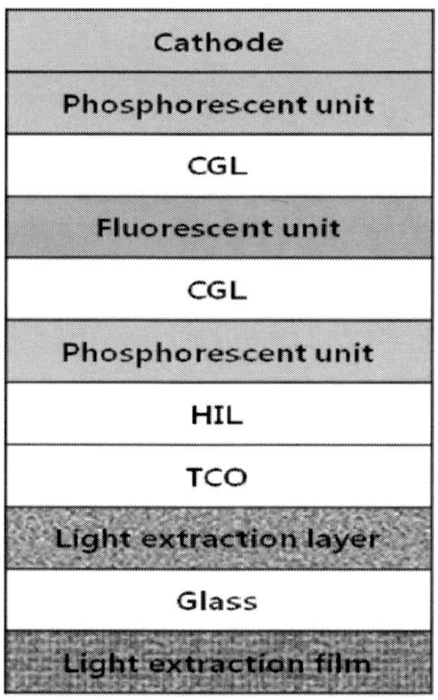

Reproduced from reference [81].

Figure 22. Schematic diagram of 3-tandem OLED lighting devices with internal and external light extraction layers fabricated by LG.

In this tandem device, one fluorescent blue emitting layer is sandwiched between two phosphorescent orange emitting layers. Between the emitting layers are charge generation layers (CGL), where LG101 is used as the hole injection material and key component for CGL. To improve the outcoupling efficiency, both the external light extraction film at the glass/air interface and the internal light extraction layer at the glass/TCO interfaces are effective. Internal light extraction layer of nano-sized particles is used to destruct the waveguide mode of the emitted light and thus enhance the light output from organic layer to the glass.

MLA type light extraction film was used to maximize the light extraction from glass to air. The device size was 100x100 mm^2 and the light emitting area was 90x90 mm^2. Auxiliary metal lines are used to reduce the IR drop within the device. After deposition organic and metal cathode layer, the devices were protected from the environment with metal film encapsulation method. The emission properties are measured at the luminance level of 3,000 cd/cm^2 and summarized in Table 6. The operating voltage was 8.1 V and the current density was 1.2 mA/cm^2. The efficacy was 105.7 lm/W at the correlated color temperature around 2800K with IEL-3. The brightness of the 100x100 mm^2 OLED lighting panel was measured at the 3,000 cd/cm^2 operation. And the expected LT_{70} was about 40,000 hrs. Figure 23 gives the EL spectra and lifetime of the fabricated tandem OLED lighting panels.

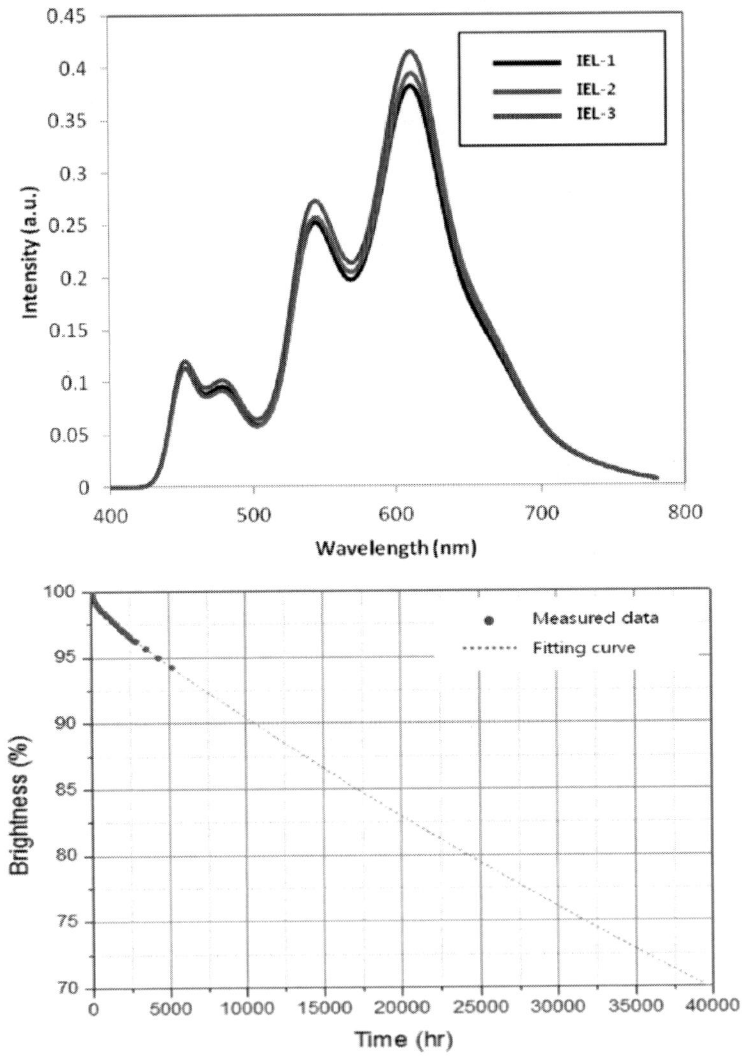

Reproduced from reference [81].

Figure 23. EL spectra and lifetime of the fabricated tandem OLED lighting panels.

Table 6. Summary of performance of the fabricated tandem OLED lighting panels

IEL structure	EQE (%)	P.E. (lm/W)	CCT	Duv	CRI
IEL-1	114	97.3	2828	0.0020	85
ILE-2	117	99.9	2785	0.0021	85
ILE-3	123	105.7	2823	0.0024	85

Reproduced from reference [81].

CONCLUSION

Tandem OLEDs are technologically interesting because not only can the luminance and current efficiency be improved linearly with the number of electroluminescent (EL) units in the tandem OLED, but also leakage current and breakdown of the electric field can be avoided due to the higher luminance at a low current density and the thicker organic films, resulting in a long lifetime. Importantly, the state-of-the-art tandem OLEDs are very easy to vertically stack either individual red, green, and blue emission units or multiple white emission units in series via charge generation layers (CGLs) to achieve white emission. In tandem OLEDs, the CGL plays a critical role in the realization of high-performance tandem OLEDs since it functions as both an internal anode and a cathode to generate intrinsic charge carriers and to facilitate opposite electrons and holes being injected into the adjacent sub-OLED units. Although amounts of CGLs have been developed to use tandem OLEDs, the problems still remain including complicated processing and limited material combination.

More importantly, since the driving voltage consumed by conventional tandem devices scales linearly with the number of electroluminescent units, the resulting power consumption would be the same for both the single-unit and tandem OLEDs to obtain the same luminescence, this means that the power efficiency cannot be increased for such tandem devices when using general CGLs. Therefore, developing new CGL structures have become important research topic in this field.

As we see, besides the luminance and current efficiency are double enhancement, the power efficiency of resulting tandem OLEDs based on organic heterojunctions as CGLs is also greatly improved. Nevertheless, the challenge posed in today's organic heterojunction-based tandem OLEDs is the fabrication of highly effective charge generation systems, that is, p-type and n-type organic semiconductors with matched energy levels and charge carrier mobility, which are always related to the operational voltage and the power efficiency.

Fortunately, these advancements have shown that the bottleneck has gradually been overcome by reasonably controlling the energy levels of p-type and n-type organic semiconductors and by introducing a proper n-doped electron-transporting layer and a p-doped hole-transporting layer to enhance the electron and hole injection, fully showing the effectiveness of organic heterojunctions as CGLs in realizing high power efficiency tandem OLEDs.

REFERENCES

[1] L. Xiao, Z. Chen, B. Qu, J. Luo, S. Kong, Q. Gong et al., "Recent progresses on materials for electrophosphorescent organic light-emitting devices," *Adv. Mater.*, vol. 23, pp. 926-52, 2011.

[2] H. Xu, R. Chen, Q. Sun, W. Lai, Q. Su, W. Huang et al., "Recent progress in metal-organic complexes for optoelectronic applications," *Chemical Society reviews*, vol. 43, pp. 3259-302, 2014.

[3] K. Leo, "Organic light-emitting diodes: efficient and flexible solution," *Nature Photonics*, vol. 5, pp. 716-718, 2011.

[4] S. Reineke, M. Thomschke, B. Luessem and K. Leo, "White organic light-emitting diodes: Status and perspective," *Reviews of Modern Physics*, vol. 85, pp. 1245-1293, 2013.

[5] J. A. Rogers, T. Someya and Y. Huang, "Materials and Mechanics for Stretchable Electronics," *Science*, vol. 327, pp. 1603-1607, 2010.

[6] C. W. Tang and S. A. VanSlyke, "Organic electroluminescent diodes," *Applied Physics Letters*, vol. 51, p. 913, 1987.

[7] C. Murawski, K. Leo and M. C. Gather, "Efficiency roll-off in organic light-emitting diodes," *Adv. Mater.*, vol. 25, pp. 6801-27, 2013.

[8] M. A. Baldo, D. F. O'Brien, Y. You, A. Shoustikov, S. Sibley, M. E. Thompson et al., "Highly efficient phosphorescent emission from organic electroluminescent devices," *Nature*, vol. 395, pp. 151-154, 1998.

[9] M. A. Baldo, C. Adachi and S. R. Forrest, "Transient analysis of organic electrophosphorescence. II. Transient analysis of triplet-triplet annihilation," *Physical Review B*, vol. 62, pp. 10967-10977, 2000.

[10] T. C. Rosenow, M. Furno, S. Reineke, S. Olthof, B. R. Lüssem and K. Leo, "Highly efficient white organic light-emitting diodes based on fluorescent blue emitters," *Journal of Applied Physics*, vol. 108, p. 113113, 2010.

[11] Y. H. Chen, J. S. Chen, D. G. Ma, D. H. Yan and L. X. Wang, "Tandem white phosphorescent organic light-emitting diodes based on interface-modified C-60/ pentacene organic heterojunction as charge generation layer," *Applied Physics Letters*, vol. 99, p. 103304, 2011.

[12] S. Lee, H. Shin and J.-J. Kim, "High-Efficiency Orange and Tandem White Organic Light-Emitting Diodes Using Phosphorescent Dyes with Horizontally Oriented Emitting Dipoles," *Advanced Materials*, vol. 26, pp. 5864-5868, 2014.

[13] J. Kido, T. Matsumoto, T. Nakada, J. Endo, K. Mori, N. Kawamura et al., "27.1: Invited Paper: High Efficiency Organic EL Devices Having Charge Generation Layers," In: *2003 SID Symposium Digest of Technical Papers*, 2003, pp. 964-965.

[14] L. S. Liao and K. P. Klubek, "Power efficiency improvement in a tandem organic light-emitting diode," *Applied Physics Letters*, vol. 92, p. 223311, 2008.

[15] L. S. Liao, W. K. Slusarek, T. K. Hatwar, M. L. Ricks and D. L. Comfort, "Tandem Organic Light-Emitting Diode using Hexaazatriphenylene Hexacarbonitrile in the Intermediate Connector," *Advanced Materials*, vol. 20, pp. 324-329, 2008.

[16] J. X. Tang, M. K. Fung, C. S. Lee and S. T. Lee, "Interface studies of intermediate connectors and their roles in tandem OLEDs," *Journal of Materials Chemistry*, vol. 20, pp. 2539-2548, 2010.
[17] Y. H. Chen and D. G. Ma, "Organic semiconductor heterojunctions as charge generation layers and their application in tandem organic light-emitting diodes for high power efficiency," *Journal of Materials Chemistry*, vol. 22, pp. 18718-18734, 2012.
[18] T.-W. Lee, T. Noh, B.-K. Choi, M.-S. Kim, D. W. Shin and J. Kido, "High-efficiency stacked white organic light-emitting diodes," *Applied Physics Letters*, vol. 92, p. 043301, 2008.
[19] Q. Y. Bao, J. P. Yang, Y. Q. Li and J. X. Tang, "Electronic structures of MoO[sub 3]-based charge generation layer for tandem organic light-emitting diodes," *Applied Physics Letters*, vol. 97, p. 063303, 2010.
[20] J. Meyer, M. Kroger, S. Hamwi, F. Gnam, T. Riedl, W. Kowalsky et al., "Charge generation layers comprising transition metal-oxide/organic interfaces: Electronic structure and charge generation mechanism," *Applied Physics Letters*, vol. 96, p. 193302, 2010.
[21] K. S. Yook, S. O. Jeon, S.-Y. Min, J. Y. Lee, H.-J. Yang, T. Noh et al., "Highly Efficient p-i-n and Tandem Organic Light-Emitting Devices Using an Air-Stable and Low-Temperature-Evaporable Metal Azide as an n-Dopant," *Advanced Functional Materials*, vol. 20, pp. 1797-1802, 2010.
[22] H. Sasabe, K. Minamoto, Y. J. Pu, M. Hirasawa and J. Kido, "Ultra high-efficiency multi-photon emission blue phosphorescent OLEDs with external quantum efficiency exceeding 40%," *Organic Electronics*, vol. 13, pp. 2615-2619, 2012.
[23] J. Xiao, X. X. Wang, H. Zhu, X. Gao, Z. H. Yang, X. H. Zhang, S. D. Wang, "Efficiency enhancement utilizing hybrid charge generation layer in tandem organic light-emitting diodes," *Applied Physics Letters*, vol. 101, p. 013301, 2012.
[24] J.-P. Yang, Q.-Y. Bao, Y. Xiao, Y.-H. Deng, Y.-Q. Li, S.-T. Lee et al., "Hybrid intermediate connector for tandem OLEDs with the combination of MoO3-based interlayer and p-type doping," *Organic Electronics*, vol. 13, pp. 2243-2249, 2012.
[25] J.-P. Yang, Y. Xiao, Y.-H. Deng, S. Duhm, N. Ueno, S.-T. Lee et al., "Electric-Field-Assisted Charge Generation and Separation Process in Transition Metal Oxide-Based Interconnectors for Tandem Organic Light-Emitting Diodes," *Advanced Functional Materials*, vol. 22, pp. 600-608, 2012.
[26] Y.-H. Deng, Q.-D. Ou, Q.-K. Wang, H.-X. Wei, Y.-Q. Li, S.-T. Lee et al., "The role of charge generation layers in the operational stability of tandem organic light-emitting diodes," *Journal of Materials Chemistry C*, vol. 2, pp. 1982-1989, 2014 2014.
[27] D.-Y. Zhou, H. Z. Siboni, Q. Wang, L.-S. Liao and H. Aziz, "The influence of charge injection from intermediate connectors on the performance of tandem organic light-emitting devices," *Journal of Applied Physics*, vol. 116, 2014.
[28] M. Y. Chan, S. L. Lai, K. M. Lau, M. K. Fung, C. S. Lee and S. T. Lee, "Influences of connecting unit architecture on the performance of tandem organic light-emitting devices," *Advanced Functional Materials*, vol. 17, pp. 2509-2514, 2007.
[29] Q.-D. Ou, L. Zhou, Y.-Q. Li, J.-D. Chen, C. Li, S. Shen et al., "Simultaneously Enhancing Color Spatial Uniformity and Operational Stability with Deterministic Quasi-periodic Nanocone Arrays for Tandem Organic Light-Emitting Diodes," *Advanced Optical Materials*, vol. 3, pp. 87-94, 2015.

[30] X. Qi, N. Li and S. R. Forrest, "Analysis of metal-oxide-based charge generation layers used in stacked organic light-emitting diodes," *Journal of Applied Physics*, vol. 107, Jan. 1 2010.

[31] M. Kroeger, S. Hamwi, J. Meyer, T. Riedl, W. Kowalsky and A. Kahn, "Role of the deep-lying electronic states of MoO(3) in the enhancement of hole-injection in organic thin films," *Applied Physics Letters*, vol. 95, 2009.

[32] B. Maennig, D. Gebeyehu, P. Simon, F. Kozlowski, A. Werner, F. Li et al., "Organic p -i- n solar cells," *Applied Physics A: Materials Science and Processing*, vol. 79, pp. 1-14, 2004.

[33] A. Yakimov and S. R. Forrest, "High photovoltage multiple-heterojunction organic solar cells incorporating interfacial metallic nanoclusters," *Applied Physics Letters*, vol. 80, pp. 1667-1669, 2002.

[34] L. S. Liao, K. P. Klubek and C. W. Tang, "High-efficiency tandem organic light-emitting diodes," *Applied Physics Letters*, vol. 84, pp. 167-169, 2004.

[35] D.-S. Leem, J.-H. Lee, J.-J. Kim and J.-W. Kang, "Highly efficient tandem p-i-n organic light-emitting diodes adopting a low temperature evaporated rhenium oxide interconnecting layer," *Applied Physics Letters*, vol. 93, 2008.

[36] M.-H. Ho, T.-M. Chen, P.-C. Yeh, S.-W. Hwang and C. H. Chen, "Highly efficient p-i-n white organic light emitting devices with tandem structure," *Applied Physics Letters*, vol. 91, 2007.

[37] M. Kroger, S. Hamwi, J. Meyer, T. Dobbertin, T. Riedl, W. Kowalsky et al., "Temperature-independent field-induced charge separation at doped organic/organic interfaces: Experimental modeling of electrical properties," *Physical Review B*, vol. 75, p. 235321, 2007.

[38] J. Birnstock, G. F. He, S. Murano, A. Wemer and O. Zeika, "Distinguished paper: White stacked OLED with 35 lm/W and 100,000 hours lifetime at 1000 cd/m(2) for display and lighting applications," *2008 Sid International Symposium, Digest of Technical Papers*, Vol. Xxxix, Books I-Iii, vol. 39, pp. 822-825, 2008.

[39] H. Nowatari, T. Ushikubo, N. Ohsawa, S. Seo and T. Tsutsui, "Intermediate connector with suppressed voltage loss for white tandem OLEDs," In: *Proc. SID Int. Symp. Digest Tech. Papers*, 2009.

[40] J. Liu, S. Huang, X. Shi, X. Wu, J. Wang and G. He, "Charge Separation Process in an Ultrathin Electron-Injecting Bilayer-Assisted Charge Generation Unit for Tandem Organic Light-Emitting Diodes," *Journal of Physical Chemistry C*, vol. 117, pp. 13887-13893, 2013.

[41] C. W. Law, K. M. Lau, M. K. Fung, M. Y. Chan, F. L. Wong, C. S. Lee et al., "Effective organic-based connection unit for stacked organic light-emitting devices," *Applied Physics Letters*, vol. 89, 2006.

[42] Y.-H. Lee, M.-W. Lin, T.-C. Wen and T.-F. Guo, "The metal interlayer in the charge generation layer of tandem organic light-emitting diodes," *Journal of Applied Physics*, vol. 114, p. 154512, 2013.

[43] S. L. Lai, M. Y. Chan, M. K. Fung, C. S. Lee and S. T. Lee, "Copper hexadecafluorophthalocyanine and copper phthalocyanine as a pure organic connecting unit in blue tandem organic light-emitting devices," *Journal of Applied Physics*, vol. 101, p. 014509, 2007.

[44] Y. H. Chen, J. S. Chen, D. G. Ma, D. H. Yan and L. X. Wang, "Effect of organic bulk heterojunction as charge generation layer on the performance of tandem organic light-emitting diodes," *Journal of Applied Physics*, vol. 110, p. 074504, 2011.
[45] Y. H. Chen, J. S. Chen, D. G. Ma, D. H. Yan, L. X. Wang and F. R. Zhu, "High power efficiency tandem organic light-emitting diodes based on bulk heterojunction organic bipolar charge generation layer," *Applied Physics Letters*, vol. 98, p. 243309, 2011.
[46] T. Chiba, Y. J. Pu, R. Miyazaki, K. Nakayama, H. Sasabe and J. Kido, "Ultra-high efficiency by multiple emission from stacked organic light-emitting devices," *Organic Electronics*, vol. 12, pp. 710-715, 2011.
[47] H. Sun, Q. Guo, D. Yang, Y. Chen, J. Chen and D. Ma, "High Efficiency Tandem Organic Light Emitting Diode Using an Organic Heterojunction as the Charge Generation Layer: An Investigation into the Charge Generation Model and Device Performance," *ACS Photonics*, vol. 2, pp. 271-279, 2015.
[48] D. Yan, H. Wang and B. Du, *Introduction to organic semiconductor heterojunctions*: John Wiley and Sons, 2010.
[49] Y. H. Chen, H. K. Tian, Y. H. Geng, J. S. Chen, D. G. Ma, D. H. Yan et al., "Organic heterojunctions as a charge generation layer in tandem organic light-emitting diodes: the effect of interfacial energy level and charge carrier mobility," *Journal of Materials Chemistry*, vol. 21, pp. 15332-15336, 2011.
[50] Q. Wang, Y. Chen, J. Chen and D. Ma, "White top-emitting organic light-emitting diodes employing tandem structure," *Applied Physics Letters*, vol. 101, p. 133302, 2012.
[51] L. Ding, Y.-Q. Sun, H. Chen, F.-S. Zu, Z.-K. Wang and L.-S. Liao, "A novel intermediate connector with improved charge generation and separation for large-area tandem white organic lighting devices," *Journal of Materials Chemistry C*, vol. 2, pp. 10403-10408, 2014.
[52] B. L. Sharma, R. K. Purohit, *Semiconductor Heterojunctions*, Pergamon Press, Oxford, 1974.
[53] Y. H. Chen, Q. Wang, J. S. Chen, D. G. Ma, D. H. Yan, L. X. Wang, "Organic semiconductor heterojunction as charge generation layer in tandem organic light-emitting diodes for high power efficiency." *Org. Electron.*, Vol. 13, p. 1121, 2012.
[54] S. Yoo, B. Domercq and B. Kippelen, "Efficient thin-film organic solar cells based on pentacene/C60 Heterojunctions," *Appl. Phys. Lett.*, Vol. 85, p. 5427, 2004.
[55] M. Kröger, S. Hamwi, J. Meyer, T. Dobbertin, T. Riedl, W. Kowalsky and H. H. Johannes, "Temperature-independent field-induced charge separation at doped organic/ organic interfaces: Experimental modeling of electrical properties," *Phys. Rev. B: Condens. Matter Mater. Phys.*, Vol. 75, p. 235321, 2007.
[56] Xiaoliang Liu, Chenggong Wang, Congcong Wang, Irfan Irfan, Yongli Gao, "Interfacial electronic structures of buffer-modified pentacene/C_{60}-based charge generation layer," *Organic Electronics*, Vol. 17, pp. 325-333, 2015.
[57] Danfeng Dai, Hongmei Zhang, Zhiqiang Zhang, Yipeng Liu, Jiangshan Chen and Dongge Ma, "Highly efficient and stable tandem organic light-emitting devices based on HAT-CN/HAT-CN:TAPC/TAPC as a charge generation layer," *J. Mater. Chem. C*, Vol. 3, p. 6809, 2015.
[58] C. Falkenberg, S. Olthof, R. Rieger, M. Baumgarten, K. Muellen, K. Leo, "The role of energy level matching in organic solar cells-Hexaazatriphenylene hexacarbonitrile as

transparent electron transport material," *Solar Energy Materials and Solar Cells*, vol. 95, pp. 927-932, 2011.

[59] C. L. Lin, T. Y. Cho, C. H. Chang and C. C. Wu, "Enhancing light outcoupling of organic light-emitting devices by locating emitters around the second antinode of the reflective metal electrode," *Applied Physics Letters*, vol. 88, p. 111106, 2006.

[60] R. Meerheim, R. Nitsche and K. Leo, "High-efficiency monochrome organic light emitting diodes employing enhanced microcavities," *Applied Physics Letters*, vol. 93, p. 043310, 2008.

[61] T. Y. Cho, C. L. Lin, C. H. Chang and C. C. Wu, "Optical characteristics of tandem and microcavity tandem organic light-emitting devices," *Journal of the Society for Information Display*, vol. 14, pp. 1063-1070, 2006.

[62] R. Meerheim, M. Furno, S. Hofmann, B. Luessem and K. Leo, "Quantification of energy loss mechanisms in organic light-emitting diodes," *Applied Physics Letters*, vol. 97, p. 253305, 2010.

[63] G. Yu, J. Gao, J. C. Hummelen, F. Wudl and A. J. Heeger, "Polymer Photovoltaic Cells: Enhanced Efficiencies via a Network of Internal Donor-Acceptor Heterojunctions," *Science*, Vol. 270, p. 1789, 1995.

[64] S. H. Park, A. Roy, S. Beaupre, S. Cho, N. Coates, J. S. Moon, D. Moses, M. Leclerc, K. Lee and A. J. Heeger, "Bulk heterojunction solar cells with internal quantum efficiency approaching 100%," *Nat. Photonics*, Vol. 3, p. 297, 2009.

[65] Z. He, C. Zhong, X. Huang, W.-Y. Wong, H. Wu, L. Chen, S. Su and Y. Cao, "Simultaneous Enhancement of Open-Circuit Voltage, Short-Circuit Current Density, and Fill Factor in Polymer Solar Cells," *Adv. Mater.*, Vol. 23, p. 4636, 2011.

[66] W. J. Zeng, K. S. Yong, Z. M. Kam, F. R. Zhu and Y. N. Li, "Effect of blend layer morphology on performance of ZnPc:C60-based photovoltaic cells," *Appl. Phys. Lett.*, Vol. 97, p. 133304, 2010.

[67] B. P. Rand, J. G. Xue, S. Uchida and S. R. Forrest, "Mixed donor-acceptor molecular heterojunctions for photovoltaic applications. I. Material properties," *J. Appl. Phys.*, Vol. 98, p. 124902, 2005.

[68] Yonghua Chen, Hongkun Tian, Yanhou Geng, Jiangshan Chen, Dongge Ma, Donghang Yan and Lixiang Wang, "Organic heterojunctions as a charge generation layer in tandem organic light-emitting diodes: the effect of interfacial energy level and charge carrier mobility," *J. Mater. Chem.*, Vol. 21, p. 15332, 2011.

[69] H. K. Tian, J. W. Shi, D. H. Yan, L. X. Wang, Y. H. Geng and F. S. Wang, "Naphthyl end-capped quarterthiophene: A simple organic semiconductor with high mobility and air stability," *Adv. Mater.*, Vol. 18, p. 2149, 2006.

[70] H. K. Tian, J. W. Shi, B. He, N. H. Hu, S. Q. Dong, D. H. Yan, J. P. Zhang, Y. H. Geng and F. S. Wang, "Naphthyl and Thionaphthyl End-Capped Oligothiophenes as Organic Semiconductors: Effect of Chain Length and End-Capping Groups," *Adv. Funct. Mater.*, Vol. 17, p. 1940, 2007.

[71] K. Walzer, B. Maennig, M. Pfeiffer and K. Leo, "Highly efficient organic devices based on electrically doped transport layers," *Chem. Rev.*, vol. 107, pp. 1233-71, 2007.

[72] S. Olthof, R. Meerheim, M. Schober and K. Leo, "Energy level alignment at the interfaces in a multilayer organic light-emitting diode structure," *Physical Review B*, vol. 79, p. 245308, Jun. 2009.

[73] R. L. Anderson, "Experiments on Ge-GaAs heterojunctions," *Solid-State Electronics*, vol. 5, pp. 341-351, 1962.
[74] R. Riben and D. L. Feucht, "Electrical Transport in Nge-Pgaas Heterojunctions," *International Journal of Electronics*, vol. 20, pp. 583-599, 1966 1966.
[75] C. Zener, "A Theory of the Electrical Breakdown of Solid Dielectrics," *Proceedings of the Royal Society A: Mathematical, Physical and Engineering Sciences*, vol. 145, pp. 523-529, 1934.
[76] K. McAfee, E. Ryder, W. Shockley and M. Sparks, "Observations of Zener Current in Germanium p-n Junctions," *Physical Review*, vol. 83, pp. 650-651, 1951.
[77] E. O. Kane, "Zener tunneling in semiconductors," *Journal of Physics and Chemistry of Solids*, vol. 12, pp. 181-188, 1960.
[78] Y. S. Tyan, Y. Q. Rao, X. F. Ren, R. Kesel, T. R. Cushman, W. J. Begley, N. Bhandari, "Tandem hybrid white OLED devices with improved light extraction," *SID 09 DIGEST*, p. 895, 2009.
[79] Y. H. Chen, J. S. Chen, D. G. Ma, D. H. Yan, L. X. Wang, "Tandem white phosphorescent organic light-emitting diodes based on interface-modified C60/ pentacene organic heterojunction as charge generation layer," *Appl. Phys. Lett.*, Vol. 99, p. 103304, 2011.
[80] Y. H. Chen, H. K. Tian, J. S. Chen, Y. H. Geng, D. H. Yan, L. X. Wang, D. G. Ma, "Highly efficient tandem white organic light-emitting diodes based upon C_{60}/NaT4 organic heterojunction as charge generation layer," *J. Mater. Chem.*, Vol. 22, p. 8492, 2012.
[81] Seongsoo Jang, Yeonkeun Lee, MinChoon Park, "OLED Lighting for General Lighting Applications," *SID 2015 DIGEST*, pp. 66, 2015.
[82] Young Hoon Son, Jung Min Lee, Byoung Yeop Kang and Jang Hyuk Kwon, "High Efficiency Three-Stack Tandem White OLEDs," *SID 2015 DIGEST*, p. 561, 2015.
[83] Yuan-Sheng Tyan, Yu-Xin Shen, Jun-Jun Peng, Lei Zhao, Chang Feng, Yun-Wu Sui, Hong Lu, Yi-Chao Wu, Hua-jin Ren, Qing-Hai Tian, Xin-Yan Gu, Gang-Yi Huang, "Efficient Tandem Hybrid White OLEDs for Solid State Lighting Applications," *SID 2014 DIGEST*, p. 675, 2014.
[84] Kazuyuki Yamae, Varutt Kittichungchit, Nobuhiro Ide, Masuyuki Ota and Takuya Komoda, Panasonic Corporation, "Realization of Outstandingly High Efficacy White OLED by Controlling Evanescent Mode and Wide Angular Incident Light," *SID 2014 DIGEST*, p. 683, 2014.
[85] Takuya Kawata, Satoshi Seo, Satoko Shitagaki, Hideko Inoue, Hiromi Seo, Ryo Hatsumi, Yuki Tamatsukuri, Tomoya Yamaguchi, Riho Kataishi, Toshiki Sasaki, Tsunenori Suzuki, Kohei Yokoyama, Hisao Ikeda, Yoshiharu Hirakata and Shunpei Yamazaki, "Highly Efficient OLED Devices with Device Architecture, for Reducing Drive Voltage," *SID 2013 DIGEST*, p. 685, 2013.
[86] www.lgoledlight.com.

In: Organic Light-Emitting Diodes (OLED)
Editor: Douglas Rivera

ISBN: 978-1-63484-001-9

Chapter 2

Synthesis and Characterization of Diphenyl Quinoline (DPQ) Derivatives for New-Fangled Green Lighting Technology

Soniya Raut[1], N. Thejokalyani[2] and S. J. Dhoble[3,*]
[1]Department of Applied Physics, Government Polytechnic, Nagpur, India
[2]Department of Applied Physics, Laxminarayan Institute of Technology, Nagpur, India
[3]Department of Physics, R.T.M.Nagpur University, Nagpur, India

Abstract

Since the preceding decade, OLEDs have become globally recognized exceptional area light source, which has enormous potential for the recent display applications and can overlay a novel path to create white light. In this regard, solid state lighting (SSL) - the new-fangled green technology has the prospective to go beyond the energy efficiencies of incandescent and fluorescent lighting and make the lighting eco - friendly and energy efficient. Such lighting can be achieved from light emitting devices based on either sublimely organics or polymeric materials and hence considerable attention has been grabbed in the scientific as well as in technological fields. Among the three primary RGB colours, the synthetic protocols and fabrication methods of green and red phosphors meet the necessary requirements. Conversely, the design and fabrication of blue phosphors and devices is still an ongoing challenge and hence research speaks volumes about the key challenges in attaining energy efficient blue light emitting organic emissive materials.

This chapter mainly deals with the low cost synthesis and characterization of amino, amino-chloro, dichloro and trichloro substituted derivatives of diphenyl quinoline (DPQ) by Friedlander condensation. The preparation of blended thin films with the synthesized complexes and binding polymers such as poly (methyl methacrylate) (PMMA) and polystyrene (PS) at different weight % are also illustrated. One of the main reasons for our synthesis and investigation of the present new conjugated polymeric compounds is to explore the molecular engineering approach as a means of the structural basis for the optimization of the desired solid-state properties. A single technique can not provide

[*] Corresponding author e-mail : sjdhoble@rediffmail.com

complete analysis of a solid and hence variety of instrumentation techniques were used for characterization of the synthesized materials. They include Fourier Transform Infrared Spectrometer (FTIR) and the X-ray diffraction (XRD) method, which are used to probe the structural properties. Spectrophotometer and spectroflurometer are used to study the optical properties by means of UV-Vis absorption and photoluminescence (PL) spectra. Thermal properties of the materials are analyzed by Thermo gravimetric Analysis (TGA) and Differential Thermal Analysis (DTA). The results reveal that the starting materials are affordable for the chemical synthesis of DPQ derivatives; they are thermally stable and offer high intensity light emission in blue region of electromagnetic spectrum. Hence, these materials can be considered as an ultimate substitute for the existing blue organic light-emitting materials, generally employed for OLEDs, display devices and many more applications in this photonic age.

Keywords: Solid state lighting, OLEDs, DPQ derivatives, Friedlander condensation, flat panel displays

1. INTRODUCTION

Conserving energy is a collective responsibility of each and every citizen for a better tomorrow. In this regard lighting technology plays a multidimensional role in the world wide development as 33% of generated electricity is consumed only for lighting. Solid state lighting – a potential green lighting with solid state devices like light emitting diodes (LEDs) and organic light emitting diodes (OLEDs) and Polymer light emitting diode (PLEDs) conserves energy, money and also offer flexibility in designs in all kinds of primary colours. Besides reducing electricity, it also helps in reducing the green house emission levels bringing down the global warming and reducing dependency on the natural resources and non-renewable energy. At present, they are little expensive than the other lighting options but they are cost effective in long run. This makes them to be more suitable for lighting applications as they possess high efficiency and long life-time. Light emitting diodes are point sources and hence need an array of diodes in order to make a display, while organic light-emitting devices are surface emission devices which can directly serve as an emitting source with design flexibility. Recently, white organic light-emitting devices (WOLEDs) have drawn much attention due to their applications as backlight for liquid–crystal displays and for solid-state lighting [1-4]. Phosphorescent OLEDs have been shown to harvest 100% of the excitons generated by electrical injection, corresponding to a four times increase in efficiency as compared to that achievable in singlet-harvesting fluorescent OLEDs [5]. For this reason, white OLEDs using phosphorescent emitter has been a focus of intense research. However, the bottleneck for further enhancement in the efficiency of white OLEDs is the performance of blue phosphorescent emitters, which is far behind those of green and yellow phosphorescent emitters. Even in flat-panel display applications the full-colour capability is required, especially for the blue one. Hence, much progress has been made in the development of blue colour OLEDs for flat panel displays. Full-colour OLEDs can be constructed in different ways, e.g., patterning pixels with the three principle colours (blue, green, and red) or with the blue and orange–yellow complementary colours, white OLEDs with colour filters [6-7], etc. Hence, it is obvious that blue light emission plays a key role in the manufacture of full colour displays [8-9]. Many materials such as distyrylarylene derivatives (DSA) [10], hydroxyphenyl–pyridine beryllium complex [11], unidentate organolithium complex [12], bistriphenylenyl [13], spirobifluorene- cored conjugated compounds [14], and anthracene derivatives [15–17] have been synthesized and used in blue OLEDs. Among these anthracene derivatives, 9,10- di-(2-naphthyl) anthracene (ADN) is one of the most promising blue fluorescent materials because of its colour purity and thermal stability[18]. Basically there are four approaches to obtain blue light emission (i) to synthesize organic dyes and polymers that are able to emit blue light, for example oxadiazole [19], distylylpyrazine [20], cyclopentadiene derivative [21], poly(alkyl-uorene) [22], poly(p-phenylene) [23], and poly-(pyridine) [24] (ii) to add blue-light emitting dyes into a polymer such as oligomeric PPV doped in polymer mixture [25] (iii) to make blue light-emitting bilayer or trilayer films from light-emitting materials with electron-transporting and hole-transporting materials like bilayer films of poly(benzoyl-1,4-phenylene) (PBP) with 2-(4-biphenylyl)-5-(4-tert-butylphenyl)-1,3,4- oxadiazole (PBD) [26] (iv) to synthesize chelate metal complexes such as azomethin zinc complex etc [27-29]. In congruent lines, we chose the synthesis of different blue light emitting diphenyl quinoline (DPQ) derivatives namely, (i) amino-DPQ (ii) amino, chloro-DPQ (iii) dichloro-DPQ (iv) trichloro-DPQ and characterization of their structural thermal and optical properties.

2. Organic Semiconductors

Organic semiconductors refer to the organic materials, which possess the ability of conducting electrons. Their electronic conductivity lies between that of metals and insulators, spanning a broad range of 10^{-9} to 10^{3} / Ω-cm. They are low molecular weight materials with fundamentally different bonding as their inorganic counterparts. In case of organic semiconductor, the bands are based on the unusual properties of the carbon atom. In organic solids, intramolecular interactions are mainly covalent, but intermolecular interactions are much weaker due to Vander Waals and London forces. As a result, the transport bands in organic crystals are much narrower than those of their inorganic counterparts and the band structure is easily disrupted by introducing disorder in the system. Thus, even in molecular crystals, the concept of allowed energy bands is of limited validity and excitations as well as interactions localized on individual molecules play a predominant role. The common electronic feature of many organic pigments is the π-conjugated system, which is formed by the overlap of carbon p_z orbitals. Due to the orbital overlap, the π-electrons are delocalized within a molecule and the energy gap between the highest occupied molecular orbital (HOMO) and the lowest unoccupied molecular orbital (LUMO) is relatively small, with transition frequencies within the visible range. By thermal activation or photo-excitation, free electrons are generated in the conduction band, leaving behind positively charged holes in the valence band. The transport of these free charge carriers is described in the quantum mechanical language of Bloch functions, k-space and dispersion relations [30].

Among other configurations, they can form sp^2 hybridizations where the sp^2 orbital form a triangle within a plane and the p_z orbitals are in the plane perpendicular to it. As bond between two carbons can be formed by an orbital overlap of two sp^2-orbitals, the energy difference between the occupied binding orbital's and the unoccupied anti-bonding orbital's is quite large, well beyond the visible spectral range. Correspondingly, longer chains of bound carbon atoms would have a large gap between HOMO and LUMO energy levels, leading to insulating properties.

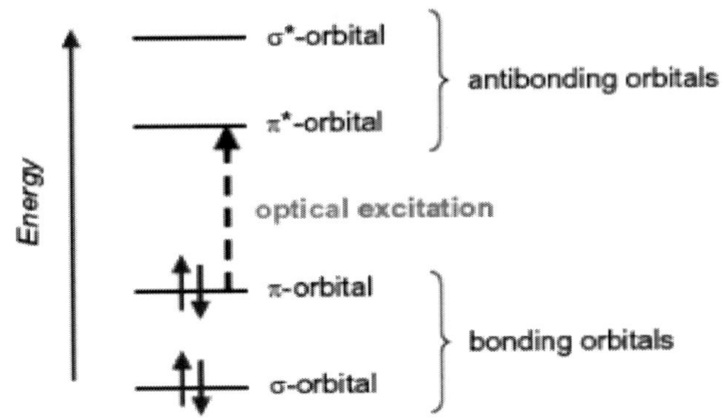

Figure 1. Energy levels of a π-conjugated molecule.

However, in sp^2 hybridization, the p_z-orbitals form additional p-bonds. These bonds have much smaller energy difference between the HOMO and LUMO leading to strong absorption in or near the visible spectral range and semiconducting properties. Organic semiconductors have a conjugated π-electron system as shown in Figure 1.The π–bonding is significantly weaker than σ-bonding and the energy gap of π–π* transitions is about 1.5 to 3 eV. However their energy gap can be controlled by the degree of conjugation in the molecule. Hence organic semiconductors have reduced hardness, lower melting point and much weaker delocalization of electronic wave functions of neighbouring molecules, which have direct effect on the optical and charge carrier transport properties.

The enormous potential application of organic materials stems from the fact that, they combine novel semiconducting properties with the scope for easy shaping and manufacture of plastics. The remarkable combination of properties made the organic semiconductors to open up new channels for electronics and optoelectronic materials. These organic semiconductors can be broadly classified into two groups on the basis of their molecular weight: conjugated polycyclic compounds of molecular weight less than 1000 and heterocyclic polymers with molecular weight greater than 1000 [31]. Polymer semiconductors are useful materials because of the ease with which they form thin films with large surface area.

However, their poor solubility in organic solvents and loss of mobility upon functionalization to enhance solubility has motivated researchers to investigate the use of small molecules as semiconductors.

Therefore, organic semiconductors can be broadly classified as

- Small-molecule organic semiconductors
- Polymer organic semiconductors

An advantage of small molecules is more facile control of charge transport by modification of various molecular parameters. For example, the ability of these molecules to pack into well organized polycrystalline films leads to higher mobility compared to polymeric semiconductors.

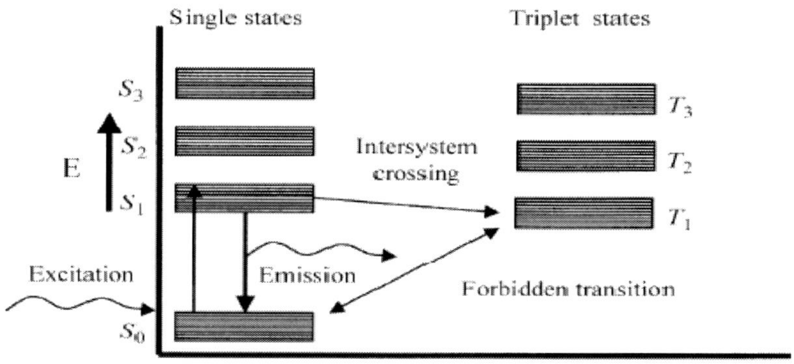

Figure 2. Jablonski diagram showing the energy levels of a hypothetical organic molecule.

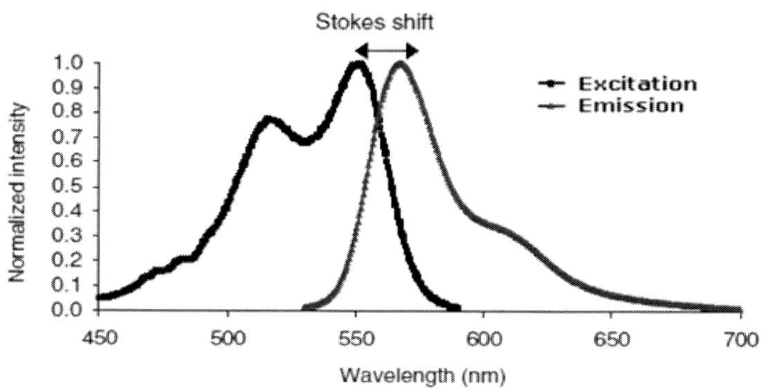

Figure 3. Molecular emission spectrum, illustrating the broadening of the spectral lines.

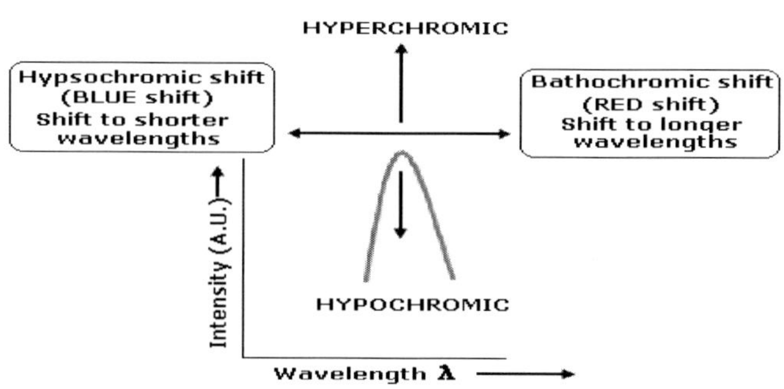

Figure 4. Terminologies denoting the absorption and intensity shift.

The polymer is used for its production and mechanical advantages without fret about optical properties [32]. Light is produced in organic materials by the fast decay of excited molecular states, and the colour of light depends on the energy difference between those excited states and the molecular ground level. Many materials show intense PL with quantum yield near about unity, while the EL efficiency is limited by the probability of creating non-radiative triplet exited states in the electron–hole recombination.

Most of the organic materials preferentially support the transport of either electrons or holes with their mobilities in the range of 10^{-8} to 10^{-2} $cm^2/(V\ s)$. Electron mobility in organic materials is generally orders of magnitude lower than hole mobility. Hence these materials are preferred as hole-transport materials (HTMs), electron-transport materials (ETMs), emitter materials and dopant materials.

Molecular orbitals of organic molecules are as shown in Figure 2. The critical difference is that by virtue of the many vibrational states of organic molecules, the energy levels become bands. This broadens the absorbance and emission spectra, as shown in Figure 3. In a typical postulated sequence, a molecule is in the lowest energy level of the ground state. It can absorb light of a diversity of wavelengths, bringing it to some vibrational level, say the first singlet excited state.

It then decays non-radiatively to the lowest level of the excited state, where it stays for about 10^{-8} to 10^{-9} sec and then decays to any of the vibrational levels of the ground state. Finally, it gives up energy non-radiatively to return to the lowest vibrational level of the ground state. In addition to broadening the spectra, these vibrational levels result in several other interesting phenomena. It is clear that the emission energy is less than that of excitation energy.

Thus, the emitted light is redder; it has a longer wavelength than the exciting light. This is known as the Stoke's shift, named after George Gabriel Stokes - the father of fluorescence.

Almost intuitively; the shape of excitation spectrum reflects the distribution of excited state vibrational levels, whereas the shape of emission spectrum reflects the distribution of ground state energy levels. Also, the vibrational levels of the triplet states can overlap the vibrational levels of the singlet states.

This affords the molecule a way of non-radiatively going from a singlet state into the corresponding triplet state.

The presence of one or more unsaturated linkages in a compound is responsible for the colour of the compound; these linkages are referred to as chromophores. Any group which does not itself act as a chromophore and its presence brings about a shift of the absorption band towards the longer wavelength. Such an absorption shift towards longer wavelength is called red shift or bathochromic shift. By the change of medium or removal of conjugation, the absorption shift towards shorter wavelength or towards blue end of the spectrum, called hypsochromic effect or blue shift. Increased conjugation usually results in increased intensity, termed hyperchromism, while decrease in intensity is termed as hypochromism. Terminologies denoting the absorption and intensity shift are shown in Figure 4.

3. Historical Development of Luminescent Materials: An Overview

Semiconductor technology is one of the most important developments in the human race and is employed in nearly all applications of high technology. One way to improve the properties of this technology is by the use of organic semiconductors. The basis for organic semiconductors as a whole revolves around the ability to develop the required material parameters, e.g., the desired LUMO and HOMO energy bands. Weuss et al., are credited with the discovery of perylene-iodine complex in 1952, the first semiconducting organic [33]. The first electroluminescence (EL) observed in organic materials followed shortly in the 1950s when acridine orange and quinacrine were subjected to high-voltage alternating current [34-37]. In 1960, researchers at the Dow Chemical company fabricated an a.c. driven EL cells using doped anthracene, a π-conjugated, three-ring, fused aromatic system. This work resulted in the first patent on EL devices based on polynuclear aromatic organic compounds [38]. Later in the decade, Pope prepared a d.c. EL cell using single crystals of anthracene, which was observed to fluoresce in the blue region by applying 400 V across the specimen [39]. The preparation techniques and latent material impurities were responsible for creating conditions requiring several hundred volts to stimulate light emission. The requirement for high operating voltages was a key factor prohibiting commercialization.

In the late 1960s and early 1970s, advances were made in organic EL device structure, two of which includes (i) the use of a reactive cathode in organic EL devices, demonstrated in 1969 [40]. This cathode structure was fabricated to permit the facile injection of electrons into the EL material, which was observed to enhance the device efficiency. (ii) In 1975 the first organic EL devices made with polymer, polyvinyl carbazole (PVK) doped with perylene, tetraphenylbutadiene or acridine orange for increased luminescence, were tested using an alkali cathode, whose work function more closely matches with the electron affinity of the EL layer [41]. The work on cathode structure by Digby, Schadt and Partridge enabled OLEDs to operate at lower voltages, a significant improvement in the general design structure. This and related early work in this field was reviewed in 1982 publication by Vincett and co-workers [40]. The development of organic luminescence progressed to the summit of practical importance through the effort of Tang in the Kodak Research Laboratories, culminating the demonstration of a two-layer structure of vacuum-deposited, small-molecule films in 1987 [43-45].

Tang and co-workers employed a transparent hole-injecting electrode of indium tin oxide (ITO) on glass to enhance the extraction of light. They sandwiched a P-type, hole-transporting molecular film of aromatic diamine and an emitting layer of the N-type, electron-transporting metal chelate tris-hydroxyquinoline aluminum (Alq_3) between the ITO anode and a low work-function alloy of magnesium as the electron-injecting cathode. Both diamine and Alq_3 possess conjugated carbon bonds and were prepared using train sublimation technique, which is generally employed to purify small-molecule compounds. Magnesium was used as cathode in order to more closely match the π* energy band of the luminescent semiconducting material to achieve electron injection with lower applied electric fields. The use of purified materials, two-layer structure and properly chosen electrodes permitted OLEDs to operate at voltages below 10 V for the first time.

Shortly, in 1990 J. Burroughs of the Cambridge group announced greenish-yellow EL device using the conjugated polymer poly(p-phenylene vinylene) (PPV) in a single-layer device structure [46]. PPV, with its aromatic phenyl ring and conjugated vinylene linkage, was first synthesized by Wessling at Dow [47-49] and has a π-π* electronic energy gap of about 2.5 eV. A precursor soluble polymer was first prepared and then a film is applied from solution by spin casting, which is then thermally converted to the conjugated form. R. Friend from Cambridge Display Technology (CDT) exploited a key patent of Cambridge University in light emission from conjugated polymers [50].

In 1991, D.Braun et al. at the University of California , Santa Barbara announced the EL application of a soluble derivative of PPV, namely poly [2-methoxy-5-(2'-ethylhexyloxy)-1,4-phenylenevinylene] or MEH-PPV [51]. It has a π-π* electronic energy gap of about 2.2 eV, which is red-shifted from that of PPV. Owing to its dialkoxy side chain, MEH-PPV offered the advantage of being soluble in conjugated form in organic solvents. However, due to its susceptibility to photo-oxidation, this material has since fallen out of favour as a candidate for commercial use [52, 53]. They also employed calcium as the reactive cathode metal in order to improve the device efficiency. In the last decade, other conjugated polymers including polythiophenes [54-56] and polyphenylenes [57] have been used for EL application. Yoshino and co-workers prepared the wide-band gap polymer polyfluorene [58] using the ferric chloride oxidative route, which was applied in a blue electroluminescent cell by Fukuda et al., in 1991 [59].

In 1992, Hosokawa et al., proposed an approach in which a nonconjugated polymer film with hole transporting ability is used as emitting layer [60]. The emitting layer was composed of a hole transporting polycarbonate containing styrylamine (SA-PC), which is easily soluble in many solvents as a functional repeating unit SA-PC. Very smooth and dense thin films of thickness less than 100 nm without any recrystallization were easily made. It seems to be very promising to realize more bright and efficient blue EL cells, since a functional repeating unit in SA-PC can be easily modified and portion of the functional repeating unit can be increased. Colour emission can be tuned by modification of the functional repeating unit. Kido et al., [61], announced that MDPs are binary solid solution of a charge transporting molecules, which is molecularly dispersed in inert, inactive polymetric binders and have been used as photoconductors in xerography. The MDPs used in this study was the molecular dispersion of N, N'-diphenyl-N,N'-bis (3-methylphenyl) - 1,1'- biphenyl-4,4'-diamine (TPD) and tris (8-quinolinolato) - aluminum(III) complex in poly (methyl methacrylate) (PMMA). Thus, the use of MDPs enables easy fabrication of EL devices in which carriers are injected through the dopants to the polymer layer leading to electroluminescence.

In 1993, polymers in the class of conjugated polyquinolines and polyanthrazolines have been synthesized and characterized by Agrawal and Jenekhe [62]. The polymers were designed to systematically vary the π-electron delocalization along the conjugated polymer backbone and, thereby to tune the electronic structure and properties in a controlled fashion. The solid polymers, which exhibit a range of light yellow to deep red colour in transmitted light were obtained in high yield with high intrinsic viscosity and high thermal stability. An investigation of the electronic structure of the polymers by optical absorption spectroscopy of the polymer thin films revealed that the structural modifications in the polymers can vary the optical band gap by about 1 eV. For example, compared to the biphenyldiyl linked polyquinoline (PBPQ) which has a band gap of 2.81 eV, the dithienylvinylene linked polyanthrazoline (PBVDA) exhibited a drastically lower band gap of 2 eV. New polymers were expected to have improved electronic, photoelectronic and nonlinear optical properties for applications and serve as model systems for investigating structure-property relationships in conjugated polymers. Yang et al., [63], reported a new macromolecule which was designed and synthesized by combining uniform conjugated units of specified length with intervening flexible-chain aliphatic oligomeric segments. The basic design concepts can be extended to produce useful soluble materials emitting in other regions of the spectrum and possessing optimized mechanical, optical and electrical properties.

In 1994 J. A. Osaheni and S. A. Jenekhe [64] explored a conjugated polymer exciplex as a new approach to enhance the solid state fluorescence quantum efficiency and blue luminescence in conjugated polymers. They presented the reports on exciplex luminescence of π-conjugated polymer which has implications and significance for understanding and developing conjugated polymers as useful optoelectronic materials, including light emitting diodes and electrophotographic photoreceptors. Kido et al., [65], observed bright blue electroluminescence from poly (N-vinylcarbazole) (PVK) multilayer-type cell, having electron transport layer. Bright light emission peaking at 410 nm can be attributed to the dimer level of the carbazole group. They also made an attempt to tune the EL colour of the PVK EL cell by the method of doping since the emission peak of the cell has the shortest wavelength as compared to the so far reported EL devices.

In 1995, Wang et al., [66] made an effort to enhance the electron-hole radiative recombination probability and carrier concentration in the emitting layer by using one polymer EL doped with another polymer, which could give rise to charge transfer. Visible reddish-orange light emission from EL diodes through poly(3-octylthiophene) (P3OT) doped with various amounts of poly(N-vinylcarbazole) (PVK) possessing hole transport character (P3OT/PVK) at low current and voltage was accomplished for the first time, which was achieved due to well-designed diode anatomy, selection and quantity control of emitting materials. Aguiar et al., [67] studied an EL polymer in which the chromophoric pendant groups are attached to a nonconjugated main chain, poly(stilbenyl-p-methoxystyrene), which was synthesized by a Williamson condensation of poly(p-acetoxystyrene) and p-(chloromethyl)stilbene.

The first reported poly (p-phenylenevinylene) (PPV), a green-blue emitter, was prepared from a soluble precursor. A new photoluminescent and electroluminescent polymer (poly-(stilbenyl-p-methoxystyrene) was prepared in which conjugation is confined to pendant chromophoric groups. The results indicate that the polymer is soluble with good film-forming ability and emits a blue light when used as the active layer of an LED. The emitting centre is assigned to an excimer formed between adjacent groups.

Voltage controlled, bicolour, light emitting electrochemical cell (LEC) made up of a polymer bilayer structure, PPV/MEH-PPV, sandwiched between two electrodes was successfully demonstrated by Yang et al., in 1996 [68]. Due to the asymmetry of electron and hole injection and transport, the dynamic light-emitting P-N junction is closer to one of the electrodes (the cathode). Therefore, when the electrode at the MEH-PPV side was wired negative, the LEC emitted the red-orange light from the MEH-PPV layer. On the other hand, when the electrode of the PPV side was wired negative, the bilayer LEC emitted the green light. The two distinct emission colours were identical to the EL colours obtained from single-layer LECs made from MEH-PPV and PPV, respectively. This indicates that the light emission of this bilayer device is completely within one polymer layer and the width of the junction is less than the thickness of that polymer layer. The bicolour LEC is the first light-emitting device which emits two completely different colours of light that does not require colour filter or a blending of polymers. The dual-colour LEC can provide a solution to simple, low cost colour displays which do not require the full range of RGB colours.

In 1996, Blom et al. [69] declared the transport properties of electrons and holes in poly(dialkoxy-p-phenylene vinylene)(PPV), investigated by current–voltage (I-V) measurements using Ca and ITO as electron and hole injecting layers, respectively. Hole transportation was well described by space-charged limited current with a hole mobility of

0.5 x 10^{-6} cm^2/V-s, while electron transportation was limited by traps, which were exponentially distributed in energy with a density of 10^{18} cm^{-3}. These results enables to further investigate the relevance of the unbalanced electron and hole transport to the device efficiency of a polymer LED. They also demonstrated that the presence of electron traps gives rise to a bias dependent efficiency. At low bias, the injected electrons were trapped and hence no contribution to the radiative recombination process. With increasing bias the traps were filled, giving rise to a strong increase in the number of free electrons, participating in recombination and hence the device efficiency might have increased.

Aromatic or heteroaromatic conjugated polymers had attracted much interest because they could be used in electroluminescent diodes or as electrochemical sensors. New aromatic poly(1,3,4-oxadiazole)s having excellent film forming properties were synthesized by Schulz et al., in 1997 [70]. The investigated polyoxadiazoles showed an emission in the range of blue to yellow light. Although, poly(arylene-1,3,4-oxadiazole)s had a combination of excellent properties, their application is rather limited due to their poor solubility in organic solvents. Low molecular weight aromatic 1,3,4-oxadiazoles were scintillators and laser dyes with high photoluminescent efficiency both in solution and in polymeric matrices. Therefore, these materials had attracted much interest as electron-injection layers in light emitting diodes. Rost et al., demonstrated ether-substituted PPV-derivatives, such as DMO-PPV, MEH-PPV and DPOP-PPV, they had received considerable attention due to their remarkable photo- and electroluminescence being combined with electron donating properties, photoconductivity and high values of third order nonlinear optical susceptibility [71]. These polymers are proved to be promising candidates for innovative electro-optical applications like electroluminescence displays, laser devices and photoconductive devices.

In 1998, Cacialli et al., [72] used a side-chain polymer based on a high-electron affinity (EA) naphthalimide moiety (PNI), to fabricate single and double-layer light-emitting diodes (LEDs) with improved efficiency in the green spectral region. Doping of the PNI with a red emitting dye molecule (4-(dicyanomethylene)-2-methyl-6-(p-dimethylaminostyryl)-4H-pyran, DCM) was also studied in order to investigate the possibility of controlling the colour of emission. Shaheen et al., [73] announced a device with excellent colour purity with an emission peak at 476 nm and a full width at half maximum of 78 nm. A thin insulating layer of LiF was inserted between the emission layer and the cathode which eliminates the need for an electron transport layer, such as Alq_3. The insertion of the LiF layer was allowed for simpler device geometry with the removal of the electron transport layer, as well as the use of a high work function aluminum cathode that was more environmentally stable than materials such as magnesium or calcium. The aim of their work was to remove the need of an electron transport layer in order to obtain better colour purity of blue emission.

In 1999, Shetty et al., [74] declared the N-type conjugated polymers by virtue of their electron transport, high electron affinity and interesting photo physical properties, this led to growing interest in electronic and optoelectronic devices. Polyquinolines are high temperature conjugated polymers with glass transitions above 200°C and decomposition temperatures above 500°C; they also possess interesting mechanical, photo responsive, photoconductive, electron transport, third-order nonlinear optical and electroluminescent properties. The EL and photo physical properties of a series of polyquinolines were investigated and used to explore the effect of molecular and supra molecular structures on the light emitting properties of polymers by Zhang et al., [75]. The polyquinolines forming ground-state aggregates had enhanced emission and nonradiative decay rates compared to those forming excimers. The problem of reducing the nonradiative rate while preserving the enhanced emission rate in such aggregates of conjugated polymers remains a challenge, however, this may be addressed through molecular and supramolecular engineering strategies. From the results, it can be inferred that the N-type polyquinolines are promising emissive materials for efficient and bright light-emitting devices whose colours can be tuned throughout the visible range of electromagnetic spectrum from blue to red.

Jung et al., in 2000 [76] announced the styrene type monomer having triphenyl ethylene (TPE) unit on the side chain, synthesized and co-polymerized with styrene monomer in order to apply to the OLEDs. Their structures were identified by FTIR spectrophotometer and mass spectrophotometer. The solid co-polymer film exhibited high fluorescence at 450 nm and the photoluminescence (PL) displayed a red-shift as compared to that of TPEwith different phases but similar luminescent properties. Even the introduction of lumophore to polystyrene chain didn't affect the PL efficiency of the lumophore itself. Thus, it can be assumed that the use of a polymer can improve the device durability without affecting the luminescent properties of the chromophore used, due to the fact that PS-TPE is more stable than TPE-dispersed PS system. Z. He [77] expressed the photophysical properties of a series of pyrazolo-[3, 4-b]-quinoline derivatives studied in solution and as dopants in several polymer matrices (polycarbonate (PC), polystyrene (PS), poly (methyl methacrylate) (PMMA) and poly (*N*-vinyl carbazole) (PVK)). In solution, these materials exhibited a high quantum yield for fluorescence at 450 nm of between 76% and 100%. The variation in substitution of the pyrazoloquinoline core of this series of materials had little effect on the photophysical properties in solution or in the polymer films.

The past two decades have witnessed the great development of OLEDs, the promising candidates for display and solid-state lighting applications through the recombination of holes and electrons; singlet and triplet excitons are generated in these devices. Kim et al., in 2002 synthesized three kinds of chromophores incorporating aromatic quinoline unit as a π-conjugated bridge in order to prepare more thermally stable nonlinear optical (NLO) chromophores than general stilbene unit [78]. The NLO poly (methylmethacrylate) copolymer, polyimides and polyester were successfully synthesized by these corresponding quinoline-based monomers. Their physical and optical properties were investigated by thermogravimetry, gel permeation chromatography, UV-Vis spectroscopy, second harmonic generation (SHG) and electro-optic (EO) measurement. All the polymers exhibited better thermal stability, however their NLO activity was a little lower than that of general stilbene-based NLO polymers.

Novel series of conjugated copolymers having oxadiazole, quinoline, quinoxaline, and phenylenecyanovinylene moieties in the main chain based on fluorene were synthesized [79] in good yields by palladium-catalyzed Suzuki coupling reaction, a new approach which differs from the traditional polyhydrazide precursor route (oxadiazole-containing polymers), acid-catalyzed Friedlander condensation reaction (polyquinolines) and Knoevenagel condensation polymerization (poly(phenylenecyanovinylene). Moreover, they introduced sterically hindered substituents into fluorine moiety to enhance processability and amorphicity of polymers. All these polymers possess excellent thermal stability with glass transition temperatures of 114-208°C and onset decomposition temperatures of 387-415°C. These copolymers possess good solubility, excellent thermal stability, low-lying LUMO energy levels and blue-emitting properties. The polymers in thin films emit strong blue luminance with narrow bandwidth upon photoexcitation. A series of new PPV oligomers containing 8-substituted quinoline, 2,2'-(arylenedivinylene)bis-8-quinoline derivatives, were designed and synthesized by Knoevenagel condensation reaction of quinaldine, 8-hydroxy or 8-methoxy-quinaldine with aromatic dialdehydes by Fushun Liang and co-workers in 2003 [80]. These PPV oligomers were characterized by ^{1}H and ^{13}C-NMR, X-ray diffraction, elemental analysis, UV-visible and fluorescence spectroscopies. The optical and photoluminescent properties study demonstrated that the emission colour of the resulting materials varies from blue to yellow and is dependent on the substituents (π-donor and π-acceptor groups) on both sides of the conjugated molecules and the aromatic core in the middle of the conjugated backbones.

Yamamoto et al., synthesized π-conjugated poly(1,10-phenanthroline-3,8-diyl), PPhen and its 5,6-dialkoxy derivatives, PPhen(5,6-OR)s, by using an organometallic polycondensation with a zero valent nickel complex [81]. They had average molecular weights of 4300-6800. PPhen had a stiff structure, as revealed by a light scattering method and exhibited a strong dichroism in UV-Vis absorption and photoluminescence. PPhen(5,6-OR)s formed an end-to-end packing assembly assisted by the side chain crystallization of the OR groups. The electron-accepting ability of PPhen is enhanced by increasing the number of the imine nitrogens, as in poly(dipyrido[3,2-c:2',3'-e]pyridazine-3,8-diyl), PDPypy. PPhen and PPhen(5,6-OR)s were susceptible to chemical and electrochemical reduction and the reduced state showed certain stability toward oxygen in air.

Ding et al., in 2005 suggested that the bis-cyclometalated iridium complexes based on quinoline ligands shows highly efficient phosphorescent properties. The synthesis,electrochemical and photophysical study of a series of bis-cyclometalated iridium(III) complexes based on quinoline ligands had been carried out [82]. These complexes were found to emit red-orange to deep red phosphorescence with high quantum yield and short lifetime. The red OLEDs with the external quantum efficiency up to 11.3% were demonstrated. This indicates the newest ideas and most up-to-date features of phosphorescence red materials for OLEDs applications. In 2005 [83] two highly efficient phosphorescent iridium metal complexes, $Ir(DPQ)_2(acac)$ and $Ir(FPQ)_2(acac)$, based on cyclometalated quinoline ligands were synthesized by Fang-Iy Wu et al. Both the complexes were suitable for use as red-emissive materials in PLEDs with CIE (Commission Internationale de L'Eclairage) chromaticity coordinates of (x=0.68, y=0.32). The phosphorescence of these Ir complexes originates from the dominant ^{3}MLCT excited state shifts to red that occur upon introducing a phenyl substituent and/or a large conjugating aromatic ring into the ligand.

The accumulation of knowledge of the intrinsic properties of organic materials, charge-carrier transport and carrier injection are two key factors that govern the performance of a device. With more understanding on the nature of carrier mobility in different organic materials, chemists succeeded in synthesizing materials with desirable properties,physicists and engineers in optimizing the device performance. S. W. Tsang and co-workers in 2006 [84], examined the feasibility of admittance spectroscopy (AS) and susceptance analysis in the determination of the charge-carrier mobility in an organic material. Admittance spectroscopy appears to be a very valuable tool in evaluating the carrier transport properties in organic semiconductors with the advantage of relatively little material consumption. Gondek et al., in 2006 synthesized a new series of pyrazoloquinoline copolymers with N-vinylcarbazole. The materials possessed high quantum efficiency (up to 1.16%) in blue spectral range (440–480 nm) [85]. The materials conceded to be promising for LEDs emitting light in the blue spectral range. It was crucial that appropriately varying the pyrazoloquinoline derivative substitution received a red spectral shift of the electroluminescent maxima possessing the enhanced quantum efficiency. A series of novel imidazole derivatives as optoelectronic materials were designed and synthesized by Zhao et al., in 2006 [86]. These organic materials have better thermal stability with better PL properties. It was confirmed that the imidazole rings can help to improve the thermal stability of the organic materials.

The optical absorption and photoluminescence were studied in the newly synthesized methoxy (MO) and carboethoxy (CE) diphenylpyrazoloquinoline (DPPQ)-derivatives by S. Calus et al., in 2007 [87]. The photoemission spectra recorded in organic solvents of different polarities were found to be highly solvatochromic. The methoxy (MO) and carboethoxy (CE) derivatives showed broad emission bands shifted to the red with the increase of solvent polarity, whereas the phenyl-methoxy derivatives exhibited the reverse solvatochromism. The methoxy and carboethoxy [DPPQ]-derivatives emitted light in the blue and green spectral range (quantum yield 0.6–0.8%). This is very imperative for efficient energy transfer in EL devices in which the methoxy or carboethoxy [DPPQ]-derivatives could be used as dopants for polymer matrix of emitting light active layers. B. M. Bahirwar et al., [88] reported the use of the methoxy substituted 2, 4-diphenyl quinoline (OMe-DPQ) to give a sharp blue light emitting organic phosphor. This polymeric compound methoxy moiety reduced the interchain aggregates, resulting good solubility in organic solvents and high thermal stability. Upon excitation of the OMe-DPQ main chain in solid state at 385 nm, the emission spectrum displays a broad band at 434 nm, with a hint of a shoulder around 468 nm. A cyclometalated iridium(III) complex containing 2-(9,9-diethylfluoren-2-yl)pyridine [Ir(Flpy)$_3$] was prepared and used in the fabrication of both yellow and white OLEDs by Yu et al., in 2008 [89]. A hole-blocking material had been used as a hole barrier layer in between different emission layers, helping the formation of the hole limitation region. The device consisted of a four-emissive-layer architecture (made of B, G, Y and R colours) sandwiched by some charge-injection and blocking layers without the need for the more sophisticated stacked design. The excitons were effectively confined by the properly interposed hole blocking material. With the proper position of a hole barrier layer and the construction of a four-emission-layer structure involving the use of [Ir(Flpy)$_3$], the resulting WOLED showed excellent device performance and stable colour even at high luminances. Highly fluorescent isomeric donor–acceptor molecules, 2-(4-phenyl-2-quinolyl)-10-methylphenothiazine (2PQMPT) and 3-(4-phenyl-2-quinolyl)-10-methylphenothiazine (3PQMPT), were synthesized in high yields (>85%) , they exhibit a factor of 2.4 disparity in EL efficiency. Two new highly emissive phenothiazine–phenylquinoline donor–acceptor molecules, 2PQMPT and 3PQMPT were synthesized and characterized by Tae Woo Kwon et al., [90]. The two isomers exhibited efficient ICT fluorescence in both dilute solution and thin films with moderate glass transition temperatures (66-73°C). The 2PQMPT isomer had an emissive excited state with a larger ICT character and a resultant lower fluorescence quantum yield than that of 3PQMPT. Green OLEDs based on the emissive 3PQMPT isomer had a factor of 2.4 higher brightness and efficiency compared to similar 2PQMPT devices. These two molecules were found to be good model systems for the study of structure–property relationships of D–A molecules in OLEDs. Haq et al., in 2009 demonstrated efficient red OLEDs based on a wide band gap material 9,10-bis(2-naphthyl)anthracene (ADN) doped with 4-(dicyano-methylene)-2-t-butyle-6-(1,1,7,7-tetramethyl-julolidyl-9-enyl)-4H-pyran (DCJTB) as a red dopant and 2,3,6,7-tetrahydro-1,1,7,7,-tetramethyl-1H,5H,11H-10 (2-benzothiazolyl) quinolizine-[9,

9a,1gh] coumarin (C545T) as an assistant dopant [81]. They obtained a device with a current efficiency of 3.6 cd/A, $CIE_{x,y}$ coordinates of (0.618, 0.373), and a peak wavelength of 620 nm at a current density of 20 mA/cm^2. This was a promising way of utilizing wide band gap material as the host to make red OLEDs, which will be useful in improving the EL performance of devices and simplifying the process of fabricating full colour OLEDs. Using co-doping and co-host system, red OLEDs with high efficiency, low driving voltage and saturated red emission were obtained. The performance of devices has been improved considerably using this co-host emitting system. The co-doping method provided a promising way of utilizing wide energy gap materials as the host to realize high efficient red emission.

Synthesis, characterization and photophysical properties of the poly {[(2, 20-bis-(4-phenylquinoline)-1,4-phenylene]-alt-phenoxy}n, a novel quinoline derived copolymer with ether linkages, were described by Simas et al., in 2009 [92]. Polymerization reaction occurred through nucleophilic aromatic substitution between an aromatic halogen and a phenol. Structural characterization was made by FTIR, NMR, DSC and TGA. The ether linkages, apart from providing better solubility compared to full conjugated quinoline copolymers, made it possible to access the photo physical properties of the quinoline moiety as a constituent of a polymer backbone, since these linkages were responsible for the confinement of the chromophoric unit. The role played by protonation in photophysical behavior was also considered, since this class of polymers is only soluble in strong acid media. Their work comprised the study of the photophysical properties of nitrogen containing polymers with confined conjugation and the role of this element in some interesting properties.

Balamurugan et al. synthesized a series of 2,9-diaryl-2,3-dihydrothieno[3,2-b] quinolines by Friedlander annulations of 5 – aryldihydro - 3(2H) – thiophenones and 2-amino benzophenones in the presence of trifluoroacetic acid in good yields under microwave irradiation at 100°C in 2010 [93]. This prompted the synthesis of thienoquinolines, with one phenyl ring and one variable aryl ring, incorporating all the encircled portions of the diarylquinoline and thienoindoles and screens them for antimycobacterial activities. Wettach et al. reported a novel method to synthesize deep blue light-emitting organic materials [94]. Various substituted low-molecular-weight triphenylene-derivatives had been prepared in a one-step procedure and were easily available on large scale and high purity. In addition, the synthesis of an oxetane functionalized, photo-crosslinkable triphenylene-based emitter material with enhanced film-forming properties was described. The low-molecular-weight emitters were vacuum-deposited, whereas the photo-crosslinkable emitter material derivative was processed from solution. The corresponding PL emission spectra exhibit λ_{emi} around 400 nm. The reasonable efforts for the chemical synthesis of the presented materials and the possibility to adjust their electronic properties by proper choice of side groups open pathways to forthcoming structure-property relationships. The affordable starting materials for the chemical synthesis of the different triphenylene emitters and its simplicity make this material class suitable for high-volume productions and consequently for low-cost OLED devices. The solvatochromic behavior of novel carbazole containing dyes obtained by the condensation of 9-ethyl-9H-carbazole-3-carboxaldehyde and 9-ethyl-9H-carbazole-3,6-dicarboxaldehyde with (1-phenylethylidene) propanedinitrile and ethyl-2-cyano-3-phenyl-2-butenoate were described by V. D. Gupta et al., [95]. DSC-TGA analysis showed that these dyes are thermally stable up to 300°C. The UV-Vis absorption and fluorescence emission spectra of the dyes were studied in solvents of differing polarity; the dyes exhibited positive solvatochromism and solvatofluorism.

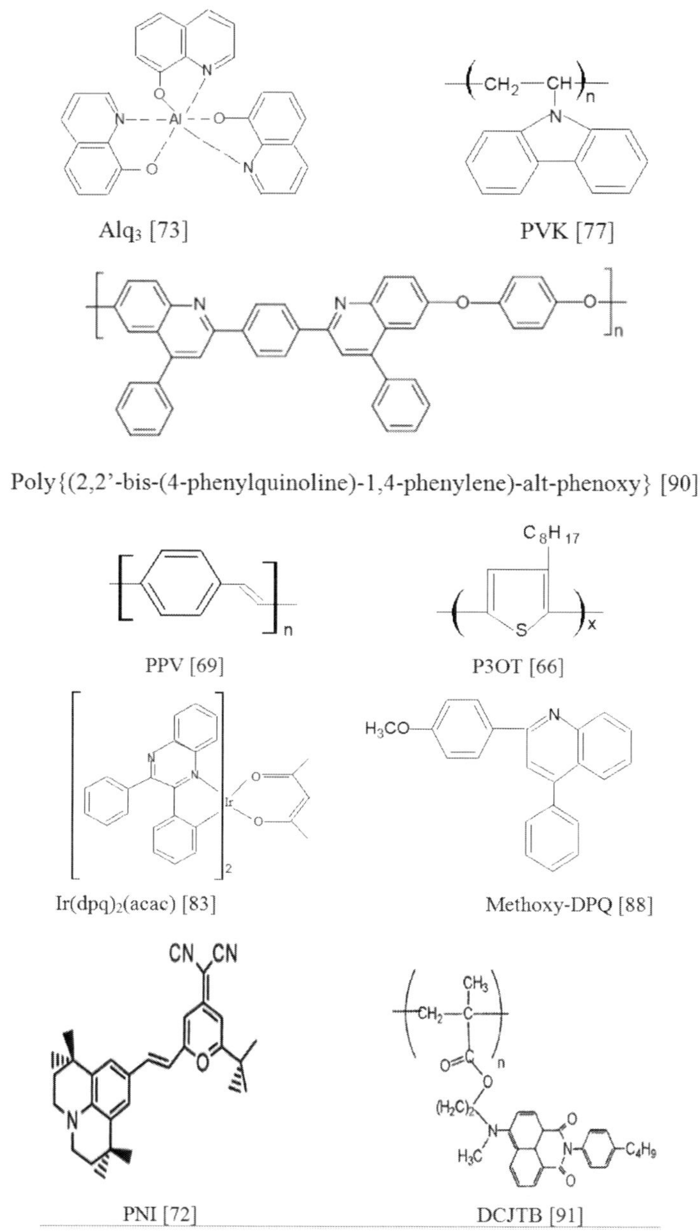

Figure 5. Chemical structures of some light emitting materials.

The solvent effect on the fluorescence characteristics of the four compounds indicates that the emission wavelength of the compounds had red shifted with the increase of solvent polarity. These dyes find applications in nonlinear optics (NLO), electronic, photonic devices and organic light emitting diodes.

Recently, new conjugated polymer, 2(4'-aminophenyl)-4-phenylquinoline i.e., amino-DPQ has been synthesized by Friedlander condensation method at 140°C by Raut et al., in 2011 [96]. The optical, structural and thermal properties of the amino-DPQ have been studied. The blended thin films of amino-DPQ with PMMA and PS at different wt% have been prepared. When powder form of amino-DPQ is excited at 385 nm, it emits intense blue light of wavelength 486 nm. The blended thin films of amino-DPQ with PMMA and PS at different wt% show the emission peak in the range of 415–450 nm. The organic compounds are blue light emitting chromophore in solid crystalline powder as well as in blended thin films. These outstanding properties make amino-DPQ, attractive candidates as advanced materials in the field of photonics and electronics. A new multifunctional compound, 4,40-di-(1-pyrenyl)-400-[2-(9,90-dimethylfluorene)]-triphenylamine (DPFA) has been designed, synthesized and applied respectively as host emitter, electron- and hole-transporters in OLEDs. S. Tao et al., announced that DPFA has efficient bipolar charge transport properties for both hole and electron [97]. The compound exhibits good thermal stability with a glass transition temperature (Tg) of 161°C and show blue emission in both solution and films. The results offer way to improve the charge balance in OLEDs by the use of emitters with bipolar charge transporting properties. Till now, reported bipolar emitters are mostly emitting in longer wavelength and there are few bipolar blue emitters.

In this digital era, most of the commentators expect improved energy efficiency and reduced energy demand to provide the dominant contribution to tackle global climate change. But at the global level, the correlation between increased wealth and increased energy consumption is very strong and the impact of policies to reduce energy demand is both limited and contested. Keeping this as essential objective, many organic light emitting materials were synthesized to use them as emissive material in the fabrication of blue OLEDs [98-107].

One of the key enablers in the history of OLED advancement can be attributed to the continuing discovery of new and improved electroluminescent materials which were made possible by the dedication and ingenuity of many organic chemists who provide the design and skilled synthesis. Indeed, from small molecules, oligomers to conjugated polymers, intense research in both academia and industry has yielded OLEDs with remarkable colour fidelity, device efficiencies and operational stability. In recent years, nanostructures of organic and complex materials have attracted a great deal of excitement due to the noncovalent intermolecular interactions such as the hydrogen bonding, Van der Waals force and π–π stacking, which have allowed them to spread among a large area used for cheap and novel optoelectronic nanodevices [108]. White OLEDs draw particular attention because of their potential applications in full colour displays, backlights, as well as general illumination purposes. White light is made up of nearly equal intensities of light from all regions (VIBGYOR) of the visible spectrum. Combining three different wavelengths of light can produce white light, provided they are widespread in the visible region. Any such sets of the three different wavelengths that on additive mixing can produce white light are called primary colours. The most common set of primary colour is red, green and blue. Another dimension to this approach is complementary colours, where the combination of only two colours can produce white light. The most used of such set is blue and yellow. Organic molecules generally have very broad emission spectra, and hence the combination of emission from not exactly complementary colours may also give white light. It is interesting to note that a single blue-emitting phosphor [109] dopant in combination with its excimer emission generated white light efficiently. The OLED have the potential to produce various colours in visible region by using a wide selection of organic fluorescent materials. So, development of organic materials for optoelectronic applications is significant. Chemical structure of some light emitting materials employed in the early devices is shown in Figure 2.

4. Organic vs Inorganic Luminescent Materials

The advantages of organic materials over inorganic materials are their excellent colour gamut and high fluorescence efficiency.

Table 1. Properties of organic versus inorganic luminescent materials

S. No.	Parameter	Inorganic	Organic
1.	Morphology	Amorphous/ Single crystalline/ Polycrystalline	Amorphous/ crystalline/ Polycrystalline
2.	Charge carrier properties	Lattice	Molecular
3.	Mobility	1-1500 cm^2/V-s	10^{-3}-10^{-5} cm^2/V-s
4.	Processing	High temperature	Low temperature
5.	Flexibility	Low	High
6.	Stability	Excellent	Currently an issue
7.	Source	Point source	Speeded source

They also offer low cost synthesis, good chemical compatibility and relative ease of handling. They possess the properties of plastics as well as semiconductors and offer simpler manufacturing process, which led to numerous applications in most of the copy machines and laser printers etc. In the mere future, number of new exciting developments like organic solar cells, organic field emission transistor and low-cost organic electronics are forefront commercial products. Inorganic versus organic material properties are shown in Table 1.

5. Quinoline: Novel Heterocyclic Precursors

Interest in the synthesis of luminescent polymeric materials continues to grow as applications in light emitting devices. Luminescent materials include small organic molecules, organometallic compounds and heteroaromatic polymers. Of the heteroaromatic polymers, quinoline based polymers are of interest as these are reported to exhibit N-type conductivity upon doping and possess excellent thermal as well as oxidative stability [110]. Most of them are heterocyclic compounds and polymers containing heterocycles. 8-Hydroxyquinoline aluminum (Alq_3) [111] and poly-p-phenylacelylene (PPV) [112] are universal OLED materials for most basic and applied research. So, quinoline and their derivatives are among the most important heterocyclic precursors [113]. For example $Almq_3$ [tris(4- methyl-8-quinolinolato) aluminum(III)] [114] and $Zn(BTZ)_2$ [bis(2-(2-hydroxy-phenyl) benzothiazole)- zinc(II)] [115] are both excellent luminescent materials and good electron transmission materials formed from quinoline derivatives coordinated with the metals Al and Zn, respectively.

The synthesis of quinolines and their derivatives has been of considerable interest to organic and medicinal chemists for many years as a large number of natural products [116] and drugs [117] contain this heterocyclic nucleus, e.g., 6-aminochrysene, discovered simultaneously in 1890 by Abegg [118]. Current research has focused on new synthetic routes to quinoline derivatives. Our choice of the phenyl quinoline (PQ) moiety as the acceptor was based on previous work that showed good electron transport properties in quinoline dendrimers [119], oligoquinolines [120] and polyquinolines [84] for OLEDs. The use of a thermally robust conjugated oligoquinoline 6, 6'-bis (2,4-diphenylquinoline) (BDPQ) as a blue-emitting and electron-transport material. The blue emitter is often doped into a wide-band-gap electron-transport host leading to high brightness and efficiency, not only due to the reduced concentration quenching in the emitter but also due to the improved electron injection and transport in the host matrix [121]. Consequently, new blue-emitting materials continue to receive considerable attention, because such devices could be utilized as a blue light source in their own right and as a host for downhill energy transfer to green or red emitting materials.

The chemical modification of the phenyl quinoline can influence the band gap and consequently change the colour of the emitted light. Quinoline derivatives are nitrogen containing heterocyclic aromatic systems with advantages of ease of synthesis and purification and used as ligands [122-123]. Methoxy substituted 2, 4-diphenyl quinoline (OMe-DPQ) [124], in this polymeric compound methoxy moiety reduced the interchain aggregates, resulting good solubility in organic solvents and high thermal stability. In the present work amino and chloro substituted DPQ derivartives were synthesized and characterized inorder to prove their ability to emit intense bright blue light in solid state as well as in PMMA and PS doped blended films.

6. Experimental

Quinoline moiety is of great importance to chemists as well as biologists as it is one of the key building elements for many naturally occurring compounds. Quinolines constitute an important class of heterocycles as the quinoline structural unit is prevalent in naturally occurring quinoline alkaloids, therapeutics and synthetic analogues with important biological activities. Several low molecular blue emitting materials such as distyrylaylenes [125, 126], metal complexes [127, 128], anthracene derivatives, oxadiazoles [129-130], lanthanide complexes [131], siloles, etc. were used for fabricating blue OLEDs. Generally, blue emitters are doped into a wide-band-gap electron–transport host. In fact, smaller concentrations reduce the concentration quenching and also improve the electron injection and transport in the host matrix. Hence, our aim is to obtain larger intensity from derivatives of DPQ and from blended thin films with both binding polymers, Poly (methyl metha-crylate) (PMMA) and Polystyrene (PS) at lower concentration. A comprehensive study of the photophysical properties of a family of eminent blue emitting derivatives of diphenyl quinoline in solution at different molar concentrations were carried out in order to study the significant spectral shift.

6.1. Synthesis of Diphenyl Quinoline Derivatives

In the present investigation, we have synthesized blue emitting *2(4'-aminophenyl)-4-phenylquinoline (amino-DPQ), 6-chloro-2(4'-aminophenyl)-4-phenylquinoline (amino, chloro-DPQ),* (2', 4'- dichlorophenyl)-4-phenylquinoline (dichloro-DPQ) and *trichloro substituted 2, 4-diphenyl quinoline (DPQ)* monomers. The blended films of these DPQ derivatives are obtained by blending them in binding polymers poly(methyl metha-crylate) (PMMA) and polystyrene (PS). These synthesized materials were characterized using FTIR, XRD, UV-Vis optical absorption and photoluminescence (PL) spectra, thermogravimetric and Differential thermal analysis.

6.1.1. Reagents and Solvents

The chemicals used for the synthesis of DPQ derivatives are :2-Aminobenzophenone, [C_{13} H_{11} NO]; FW = 197.24, m. p. = 105 to 108^{0}C (Aldrich Chemicals), 4-Aminoacetophenone, [C_8 H_9NO]; FW = 137.17, m. p. = 105 to 107^{0}C (Aldrich Chemicals), 2-Amino, 5-Chlorobenzophenone, [$C_{13}H_{10}$ClNO]; FW = 231.68, m. p. = 96 to 100^{0}C (Aldrich Chemicals), Diphenyl Phosphate, [C_{12} H_{11} PO_4], FW =250.19, m. p. = 67- 68^{0}C (Aldrich Chemicals), m- Cresol, FW = 108.14 (Alfa Aesar Chemicals), Sodium Hydroxide [NaOH], Dichloromethane [CH_2Cl_2] SRL grade, 99.5%, MW = 83.93, N-Hexane and Magnesium sulphate [$MgSO_4$] and 2', 4'-Dichloroacetophenone, [$C_8H_6Cl_2O$]; FW = 189.04, m. p. = 33 to 34^{0}C (Aldrich Chemicals).

6.1.2. Scheme of Synthesis

Synthesis of *2(4'-aminophenyl)-4-phenylquinoline (amino-DPQ) is shown in Figure 6. Initially,* 2-aminobenzophenone (2 g) and 4-aminoacetophenone (2 g) were added along with 2 g of diphenyl phosphate and 3 ml of m-cresol in a glass reactor fitted with mechanical stirrer. The reaction mixture was heated at 140°C for 4 h. After cooling, methylene chloride (100 ml) and 10% NaOH (100 ml) were added to reaction mixture. An organic layer is formed as shown in Figure 7. This layer was separated and washed with distilled water (50 ml x 5 times). Then it was dried over $MgSO_4$ and evaporated under the natural environment to yield an off-white solid. The crude product was then washed with hexane (25 ml x 5 times) to afford crystalline solid (2.65 g) (296.37, [$C_{21}H_{16}N_2$]). The presence of an amino group strongly increases the reactivity of the aromatic ring, due to an electron donating effect.

In case of amino, chloro-DPQ, amino, chloro substituted 2, 4-diphenylquinoline (DPQ) i.e., 6-chloro-2(4'-aminophenyl)-4-phenylquinoline (amino, chloro-DPQ) was synthesized according to the molecular formula. 2-amino, 5-chlorobenzophenone reacts with 4-aminoacetophenone in presence of diphenyl phosphate and m-Cresol at 140 oC, water vapours come out and finally undergoes crystallization and gives 6-chloro-2(4'-aminophenyl)-4-phenylquinoline (amino, chloro-DPQ) (330.81, [C21H15N2Cl]). Chemical structure of amino, chloro-DPQ is shown in Figure 8.

Similarly, Dichloro substituted 2, 4-diphenylquinoline (DPQ) i.e., (2', 4'-dichlorophenyl)-4-phenylquinoline (dichloro-DPQ) and *trichloro substituted 2, 4-diphenyl quinoline (DPQ)* i.e., 6-chloro-2(2', 4'-dichlorophenyl)-4-phenylquinoline (trichloro-DPQ) *were synthesized with the help of Friedlander condensation method. Their structure is shown in Figure 9 and 10, respectively.*

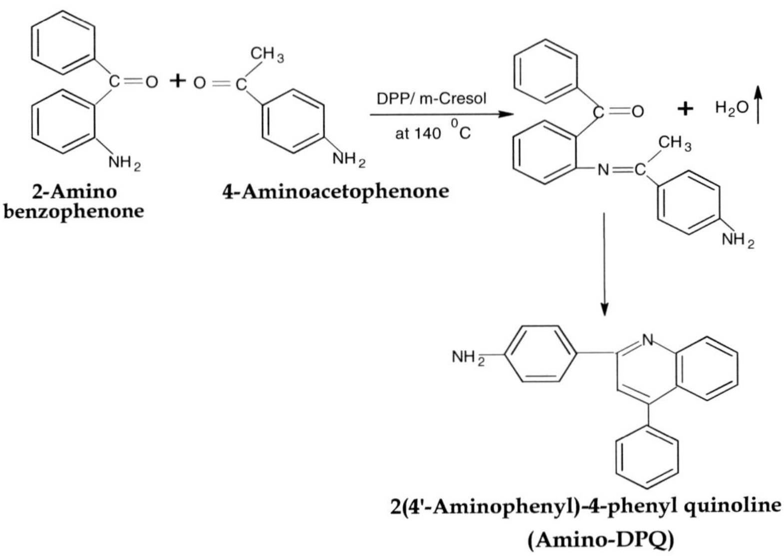

Figure 6. Synthesis of *2(4'-aminophenyl)-4-phenylquinoline (amino-DPQ).*

Figure 7. An organic layer of synthesized *amino-DPQ*.

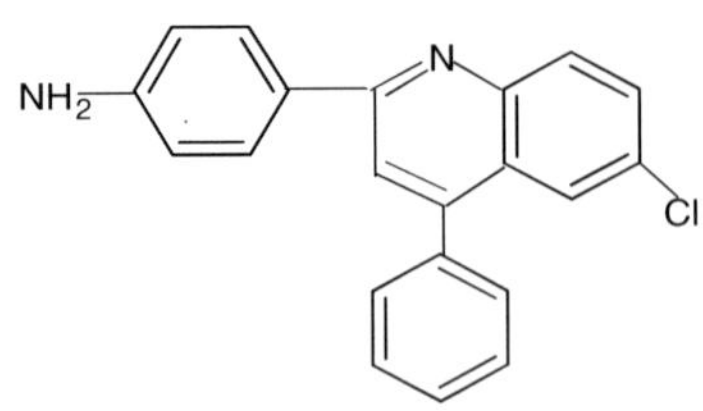

Figure 8. Chemical structure of amino, chloro-DPQ.

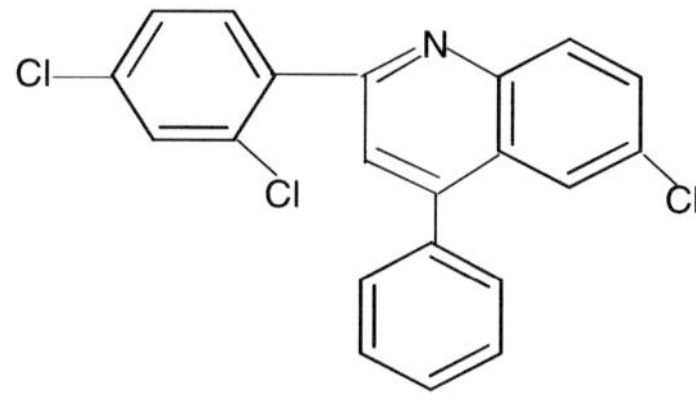

Figure 9. Chemical structure of dichloro-DPQ.

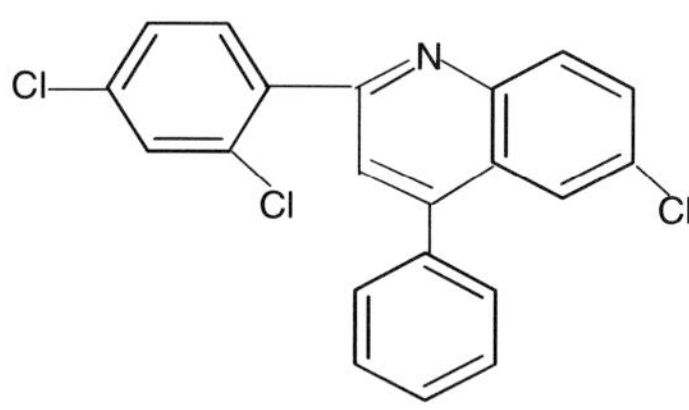

Figure 10. Chemical structure of trichloro-DPQ.

6.2. Characterization of DPQ Derivatives

6.2.1. Fourier Transform Infrared Spectrometer (FTIR)

The molecular structure of the synthesized polymeric compounds has been confirmed by FTIR. The FTIR spectra of different derivatives of diphenyl quinoline, namely, (i) amino-DPQ (ii) chloro-DPQ (iii) dichloro-DPQ (iv) trichloro-DPQ are shown in Figure 11.

The new strong bands between 1500 and 1400cm^{-1} due to the imine (C=N) group, which is the characteristic of the quinoline ring in both the compounds are observed. The C=N structure in the polymers gives the electron transporting characteristics. The N–H stretch appears at 3451–3200cm^{-1}. In amino, chloro-DPQ, a band in the region 750-700 cm^{-1} shows mono-substituted (Cl) benzene. C-N stretching absorption occurs at 1360-1250 cm^{-1}. Aromatic C-C stretch bands for the carbon- carbon bonds in the aromatic ring at about 1500-1600 cm^{-1} were observed.

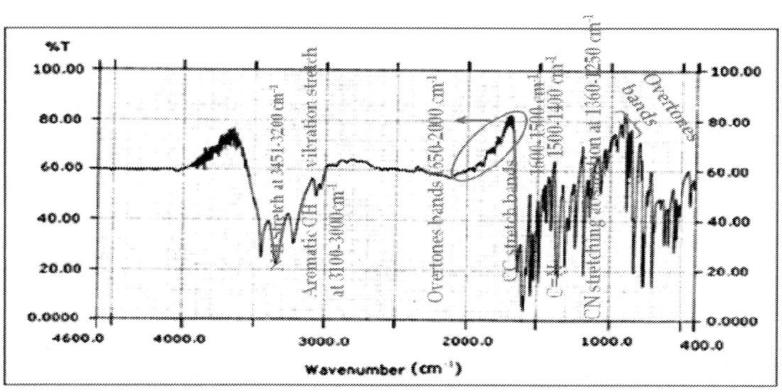

(i)

Figure 11. (Continued).

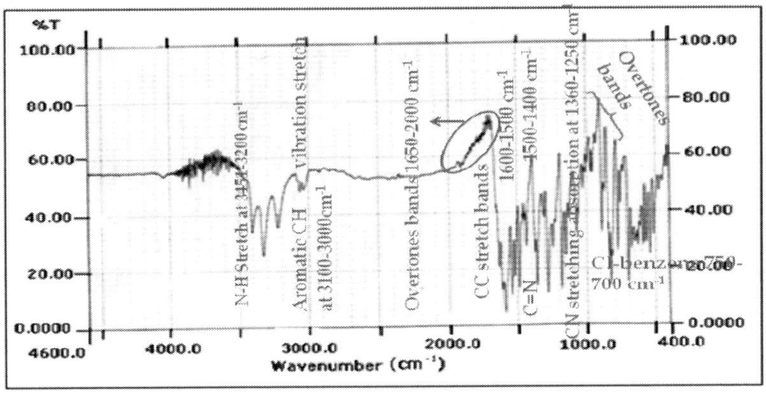

(ii)

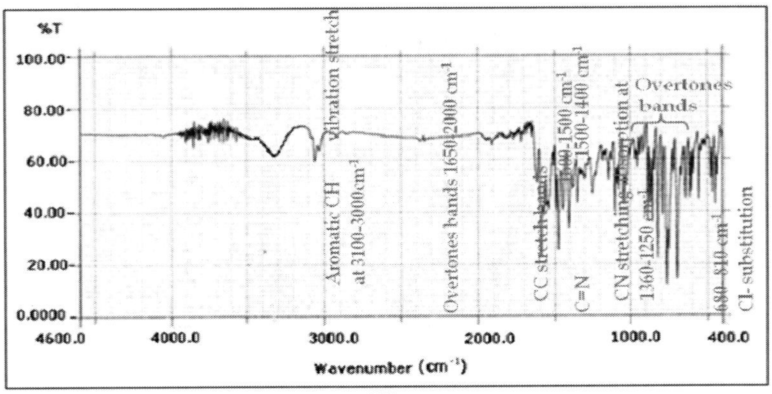

(iii)

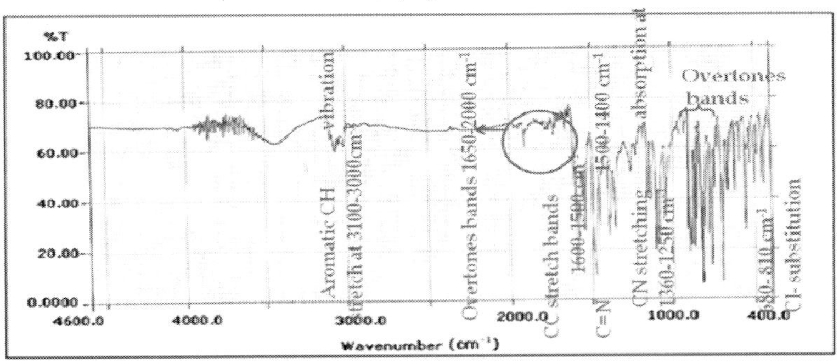

(iv)

Figure 11. FTIR spectra of (i)amino-DPQ (ii)chloro-DPQ (iii) dichloro-DPQ (iv) trichloro-DPQ.

The substitution pattern in an aromatic compound is identified by looking at the C-H out of plane bending bands and also from the combination bands (overtones) between 1650-2000 cm^{-1}. A strong band in the region 750-700 cm^{-1} shows chlorobenzene substitution. Dichloro and trichloro DPQ also display strong bands between 1500 and 1400cm^{-1} due to the imine (C=N) group which is the characteristic of the quinoline ring in both the compounds.

6.2.2. X-ray Diffraction

The diffractograms of all the synthesized diphenyl quinoline derivatives, namely amino-DPQ, amino, chloro-DPQ, dichloro-DPQ and tri-chloro-DPQ are shown in Figure 12(i), (ii), (iii) and (iv) respectively.

All these diffractograms display many strong, sharp diffraction peaks inferring their crystalline nature. These spacing corresponds to the chain distances of a well-organized molecular layer structure. Sharp narrow diffraction peaks indicates the crystalline parts, while broad peaks indicate the amorphous nature.

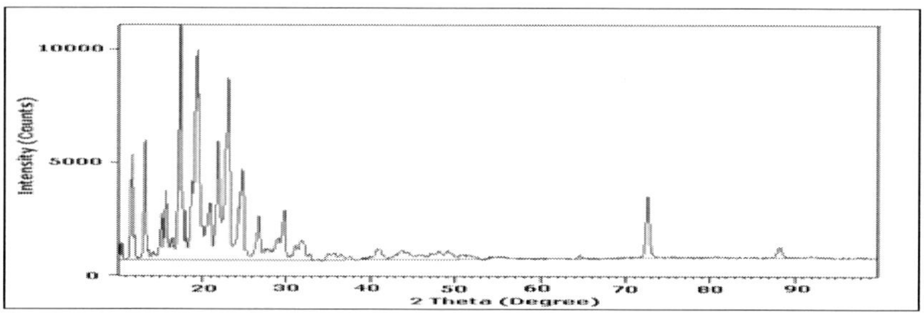

(i)

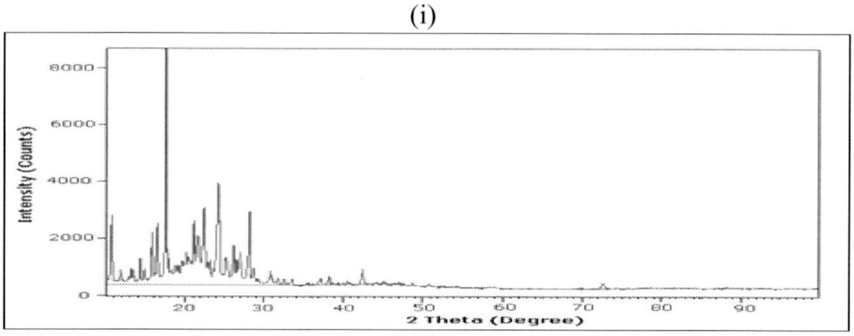

(ii)

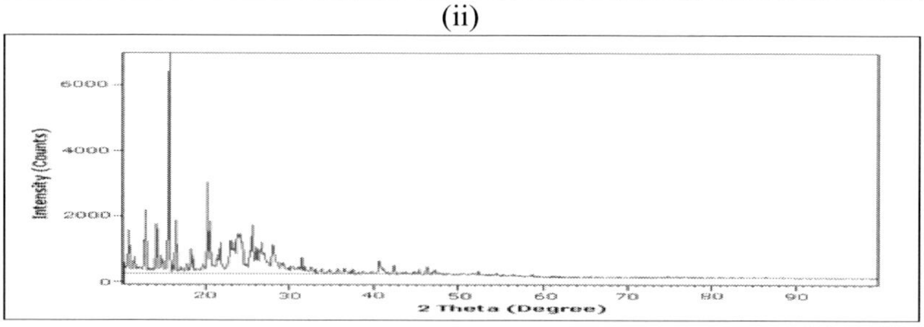

(iii)

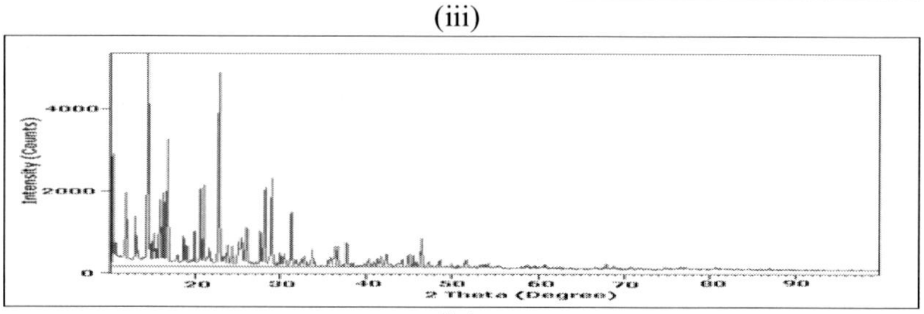

(iv)

Figure 12. XRD of (i) amino-DPQ (ii) chloro-DPQ (iii) dichloro-DPQ (iv) trichloro-DPQ.

6.2.3. TGA/DTA Measurements

Thermal stability, chemical reactivity and phase transitions properties of materials are evaluated by Thermo gravimetric analysis (TGA) and Differential thermal analysis (DTA). TGA is used for precise measurement of specimen weight changes as function of time and temperature in various gaseous environments. Figure 13 and 14 show the differential thermal analysis (DTA) and thermo gravimetric analysis (TGA) plots of the amino-DPQ and amino, chloro-DPQ, respectively. TGA and DTA measurements can be made simultaneously on the sample to allow direct correlation of weight and heat change events with greater certainty than available with separate measurements. In both the cases, TGA results reveal that the thermal stability of the polymeric compounds is around 302.80°C and 540.16°C with end peak at 354.03°C and peak at 625.59°C in amino-DPQ and amino, chloro-DPQ, respectively which indicates the stability of the material. Two endothermic peaks at 76.26°C and 155.96°C, corresponds to distortion of water from the powders and no thermal weight loss is observed in TGA curve. Around 355.38°C, a weak exothermic peak is observed in the DTA curve, which can be attributed to the decomposition process of the residual organic materials.

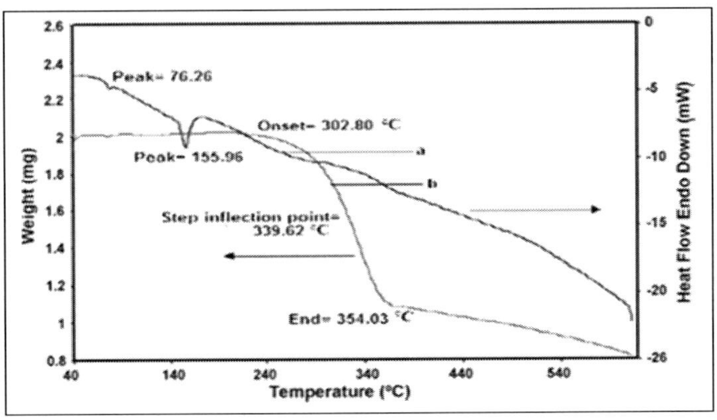

Figure 13. DTA (a) and TGA (b) curve of amino-DPQ.

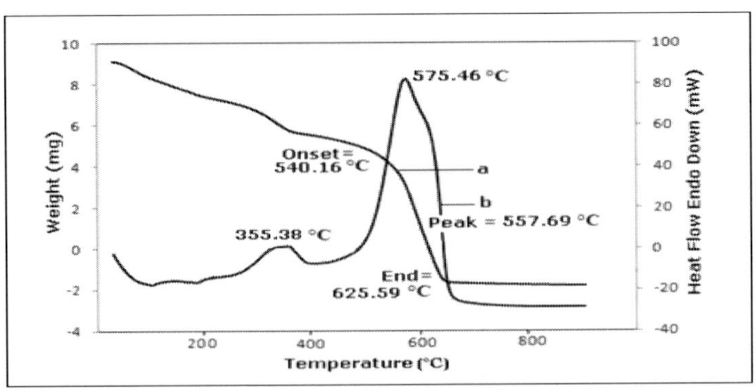

Figure 14. TGA (a) and DTA (b) curve of amino, chloro-DPQ.

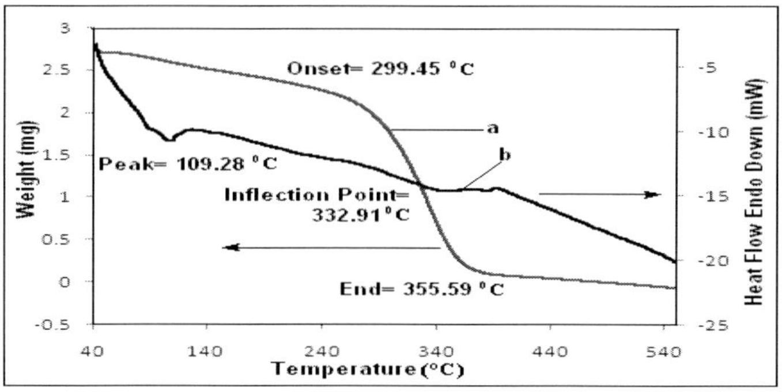

Figure 15. TGA (a), DTA (b) curves of dichloro-DPQ in powder form.

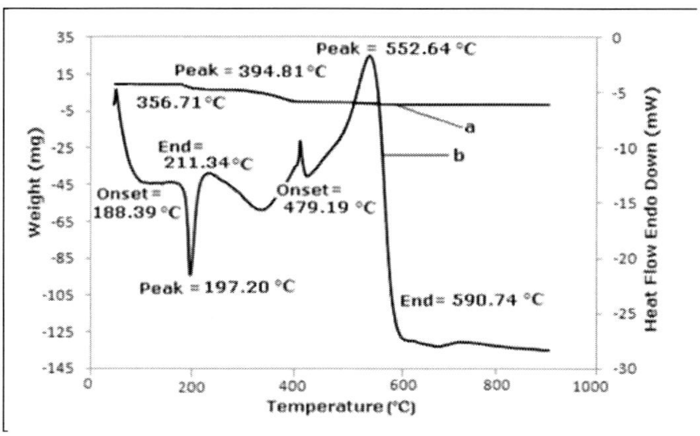

Figure 16. TGA (a), DTA (b) curves of trichloro-DPQ in powder form.

As the temperature increases, a sharp exothermic peak is observed at 575.46°C showing the phase transition. An increase in physical stability is also observed with aromaticity, due to the electronic delocalization in the π orbitals of the aromatic ring that causes an additional intermolecular interaction. This increase also occurs when halogens (compounds with higher polarity and more intense dipole–dipole interactions) are present. Evaluation of thermal stability, chemical reactivity and phase transitions properties of dichloro-DPQ and tricloro-DPQ are carried out by TGA and DTA measurements, the corresponding curves are as shown in Figure 15 and 16, respectively. DTA of dichloro substituted DPQ shows the endothermic peak temperature at about 109.28°C having onset temperature 97.03°C. The TGA result reveals that the thermal stability of the polymeric compounds is around 299.45°C, indicating that the materials are stable upto 299.45°C. The 10 mass% loss temperatures for the dichloro-DPQ was found to be 355.59°C. From this we can conclude that in the presence of the dichloro group thermal stability of the DPQ is moderately improved. In trichloro substituted DPQ, the weight loss 8.78% step is observed between 20 and 250°C, accompanied by endothermic peak at 197.20°C in the DTA curve due to the evaporation of residual moisture. Negligible weight loss occurred at 424°C and an exothermic peak was observed at 552.64°C. No weight loss was observed beyond 600°C. The TGA curve shows the good thermal stability with very little amount of weight loss. The length of the chain and/or the branching result is an increase in the stability. The presence of halogens gives rise to products with high temperatures and enthalpies of fusion, indicating that they are more physically stable. The thermal studies of the compounds show good thermal stability. The derivatives of diphenyl quinoline (DPQ) have intense light emitting properties as well as good thermal stability. Hence, it can be concluded that the starting materials are affordable for the chemical synthesis of the derivatives of DPQ, making them suitable for low-cost organic light-emitting devices and lighting applications, which play an important role in this photonic age.

6.2.4. Photoluminescence Spectra

6.2.4.1. Photoluminescence Spectra of Amino-DPQ

The excitation and photoluminescence (PL) spectra of amino-DPQ powder is as shown in Figure 17. When the sample is excited at 385 nm, it emits intense bright blue emission peaking at λ_{max} of 486 nm, which falls in the blue region of electromagnetic spectrum. Such emission is eco-friendly as it does not emit any hazardous gases. Hence these materials can be used for the fabrication of OLEDs for solid state lighting.

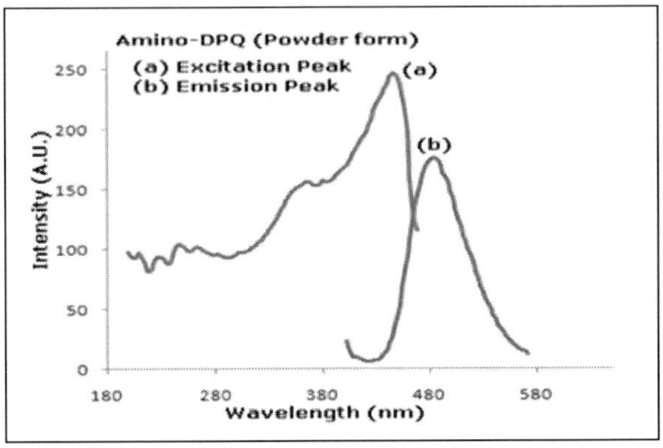

Figure 17. PL spectra of amino-DPQ (powder).

6.2.4.2. Photoluminescence Spectra of Amino Chloro-DPQ

When the amino chloro-DPQ powder is excited at 382 nm, it emits intense blue light of wavelength 479 nm as shown in Figure 18.

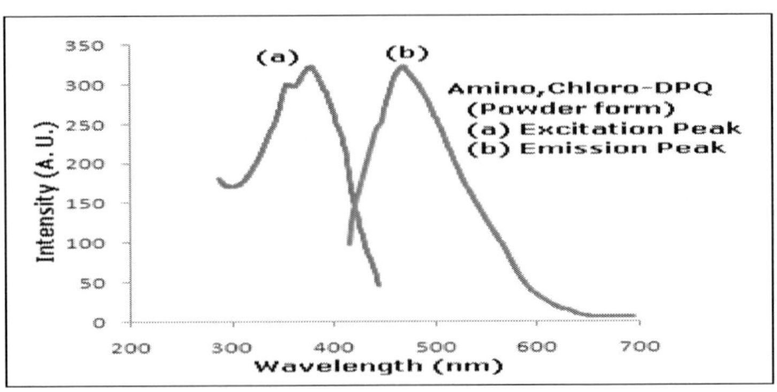

Figure 18. PL spectra of amino, chloro-DPQ (powder).

6.2.4.3. Photoluminescence Spectra of Dichloro-DPQ

The excitation spectra of dichloro-DPQ in crystalline powder display a peak at 370 nm with intense blue light emission at 462 nm as shown in Figure 19. In case of diphenyl quinoline (DPQ) in powder form, the PL spectrum shows emission at 434 nm when excited at 365 nm. Hence red shift of 28 nm is observed in the emission spectra of dichloro-DPQ in powder form, which may be due to dichloro group substitution as compared to 2, 4-DPQ and also due to dominant intramolecular charge transfer.

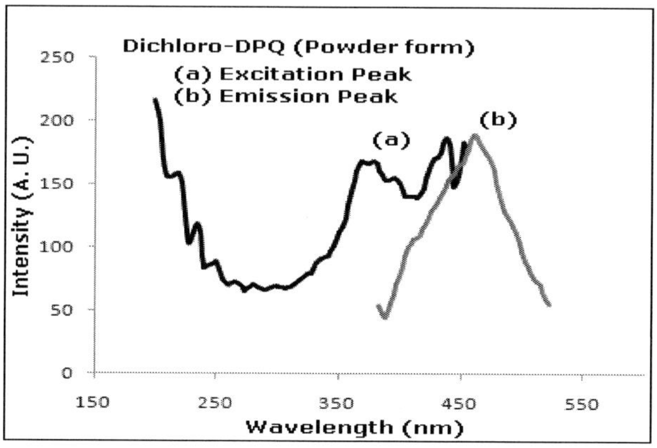

Figure 19. PL spectra of dichloro-DPQ (powder).

6.2.4.4. Photoluminescence Spectra of Trichloro-DPQ

When the powder of trichloro-DPQ is excited at 350 nm, it emits intense blue light of wavelength 460 nm as shown in Figure 20. Hence, it is generally true that the emission spectrum has much better resolved structure than in the absorption spectrum. Structural, thermal and optical properties of DPQ derivatives are tabulated in Table 2.

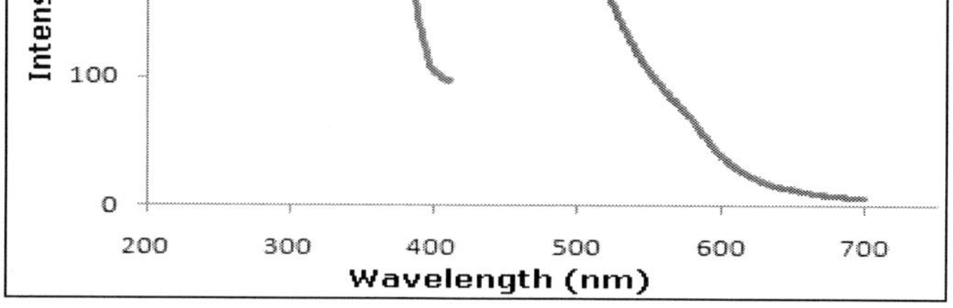

Figure 20. PL spectra of trichloro-DPQ (powder).

Table 2. Structural, thermal and optical properties of DPQ derivatives

Parameter	Amino DPQ	Amino,Cloro-DPQ	Dichloro-DPQ	Trichloro-DPQ
Molecular formula	$C_{21}H_{16}N_2$	$C_{21}H_{15}N_2Cl$	$C_{21}H_{13}NC_{12}$	$C_{21}H_{12}NC_{13}$
Appearance	Off-white Solid	Off-white Solid	Off-white Solid	Off-white Solid
Molecular weight	296.37 gm	330.81gm	350.24gm	384.69gm
Nature	Crystalline	Crystalline	Crystalline	Crystalline
d-values* (A^0)	5.10	5.05	5.67	6.16
Angle 2θ* (A^0)	17.3	17.5	15.6	14.3
Stability of complex (^{0}C)	302.80	540.16	299.45	394.81(less weight loss occurred)
Melting transition temperature (^{0}C)	140	163.50	175	194.78
Excitation wavelength (nm)	385	382	370	350
Emission wavelength (nm)	486	479	462	460
Emission intensity (a.u)	180	330	200	350

*100% relative intensity.

7. Molecularly Doped Polymers

The science and technology of polymeric mixtures has attracted much academic and industrial interest over the last few decades. As a result, the basic rules of blending are now fairly established and numerous advances have been made to underpin the role of molecular interactions on the phase behaviour of blends made from high molecular weight polymers. An important class of disordered molecular solids are molecularly doped polymers(MDP) and glassy charge-transporting materials. MDP are amorphous solid solutions of molecules of electron or hole-transporting materials (electron acceptors or electron donors, respectively) dissolved in a polymer hosts acting as a binder.

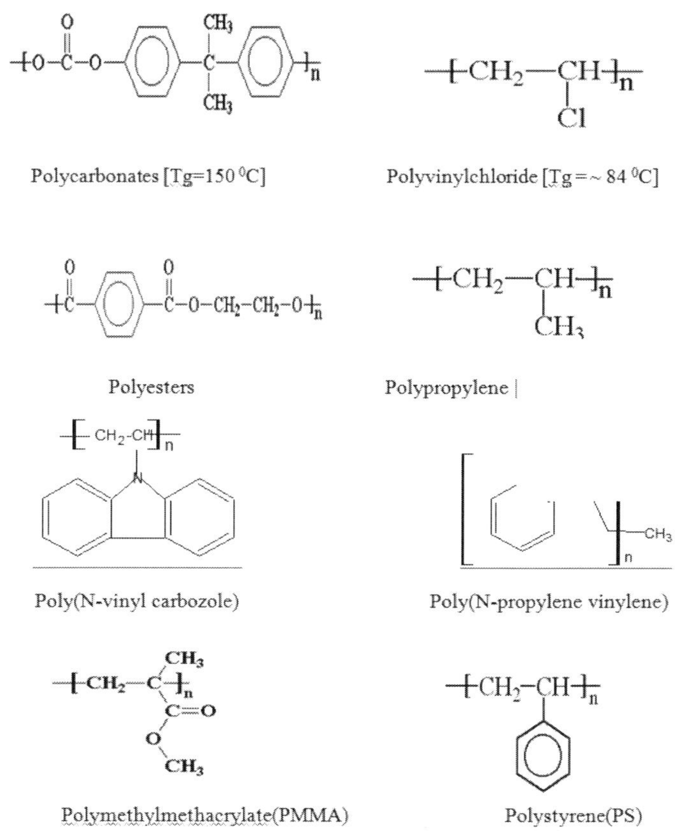

Figure 21. Most commonly used polymers for molecular doping.

Table 3. Physical and chemical parameters of PMMA and PS[136]

Chemical formula	$(C_5O_2H_8)_n$	$(C_8H_8)_n$
Synonyms	Polymethylmethacrylate PMMA poly (methyl methacrylate)	Polystyrene PS
Molecular mass	varies	varies
Density	1.19 g/cm³	1.05 g/cm³
Melting point	180 ^{0}C	237.5 ^{0}C
Boiling point	200.0 ^{0}C	-
Glass transition	114 ^{0}C	100 ^{0}C
Refractive index	1.492 (λ=589.3 nm)	1.519(λ=589.3 nm)
Transmission	80-90%	-

Charge carrier transport phenomena in molecularly doped polymers (MDPs) are of great technological and fundamental interest due to their current application as charge transport layers in organic electrophotographic photoreceptors, electroluminescent [132] and photovoltaic [133] devices. MDPs are binary solid solution of a charge transporting molecules, molecularly dispersed in an inert, inactive polymetric binders, and finds application as photoconductors in xerography [134]. Compared with intrinsically conductive polymers, MDPs are much more flexible in material design.

In this regard, PMMA (polymethyl methacrylate) is a versatile polymeric material that is well suited for many imaging and non-imaging microelectronic applications [135]. PMMA is most commonly used as a high resolution positive resist for direct write e-beam as well as X-ray and deep UV microlithographic processes. It is also used as a protective coating for wafer thinning, as a bonding adhesive and as a sacrificial layer. In these blends conventional polymers with excellent optical and mechanical properties like poly (methyl metha-crylate) (PMMA), polystyrene (PS), polyethylene (PE), polyvinylchloride (PVC), or polycarbonate (PC) can be used as host matrices. The guest– host approach is a sound method to improve the sample quality to enable the investigation of one dimensional system. Most commonly used polymers for molecular doping are shown in Figure 21.

The physical and chemical parameters of PMMA and PS were shown in Table 3.

7.1. Preparation and Characterization of Blended Films

Commercially available polymers PMMA/PS were used for making blended films of the synthesized complexes. Polymer (PMMA/PS) matrix is prepared by dissolving 0.5 gm of polymer (PMMA/PS) in 15 ml of chloroform with vigorous stirring for 15 minutes at room temperature. Later 0.05 gm of the synthesized complexes were taken individually and dissolved in the same solvent in separate beakers. Later, the complex solution was mixed with PMMA/PS matrix at room temperature under vigorous stirring for 15 minutes to obtain the homogeneous mixture. The resulting homogeneous mixture is then poured on to a glass.

The solvent is allowed to evaporate in air for 2 to 3 hrs at room temperature and then pilled up from the substrate. The obtained films are placed in vacuum dry oven at room temperature overnight to remove residual solvent left, if any. The obtained blended films are homogeneous and showed excellent optical transparency. No visible phase separation is detected. Absorption and photo luminescence measurements were recorded on SPECORD 50 spectrophotometer and Humamatsu F-4500 spectrofluorometer, respectively.

7.2. Optical Characterization of Blended Films

In order to examine the effect of doping concentration of DPQ derivatives on the optical aborption and emission intensity, four different amounts of wt% 0.1, 1, 5 and 10% of DPQ derivatives are introduced into solid matrix of PMMA.

The UV-Vis absorption spectra measurements give better insight of the absorption photon energy and excitation of molecular orbitals. It also provides a good deal of information on the electronic structures of the polymeric compound.

7.2.1. Absorption Spectra of Amino-DPQ in PMMA

The optical absorption spectra of amino-DPQ with PMMA with variation in concentration are shown in Figure 22. It displays two broad absorption bands, one peaking at 268 nm and second peaking in the range of 360–400 nm. The peak at lower wavelength is attributed to $\pi \rightarrow \pi^*$ transition while the peak at higher wavelength is attributed to $n \rightarrow \pi^*$ transitions in various molecular orbitals of amino-DPQ.

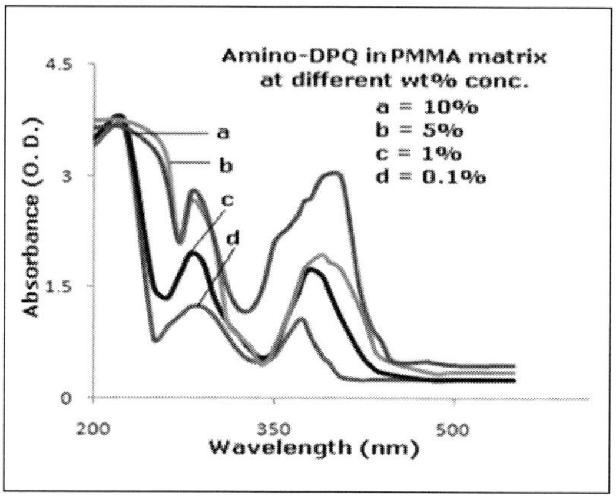

Figure 22. Absorption spectra of amino-DPQ with PMMA at different wt% .

The polymeric compound has nearly identical absorption maxima (λ_{max}) at 268 nm, while the second absorption peak shows a small shift in its position with change in doping concentration. This may be due to the substituent's which have definite effects on the absorption bands.

7.2.2. Absorption Spectra of Amino-DPQ in PS

When amino-DPQ is molecularly doped in polystyrene, well resolved absorption peaks are observed in optical absorption spectra as shown in Figure 23.

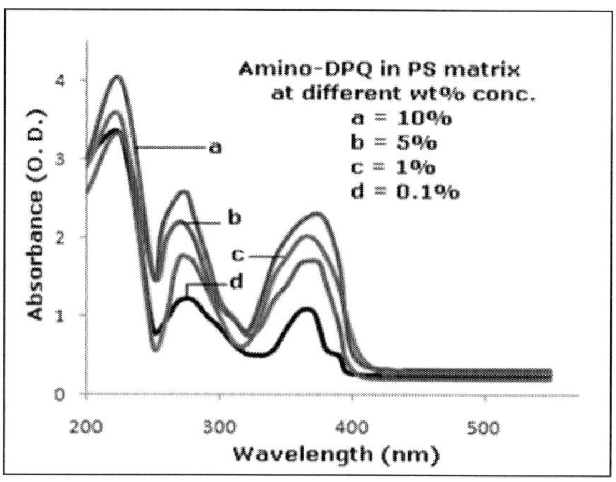

Figure 23. Absorption spectra of amino-DPQ with PS at different wt%.

Blended films of amino-DPQ showed λ_{max} at 268 nm and in between 360-370 nm. Slight change in λ_{max} was observed when the results are compared with the blended films of PMMA. In general, more the interaction between the host and the complex more will be the emission intensity. So, it is expected that the interaction the complex and the host polymer is less in polystyrene than in PMMA.

7.2.3. Absorption Spectra of Amino, Chloro-DPQ in PMMA

The optical absorption spectra of amino, chloro-DPQ in PMMA for different wt% are shown in Figure 24. It displays two broad absorption bands at 276 nm and 378 nm.

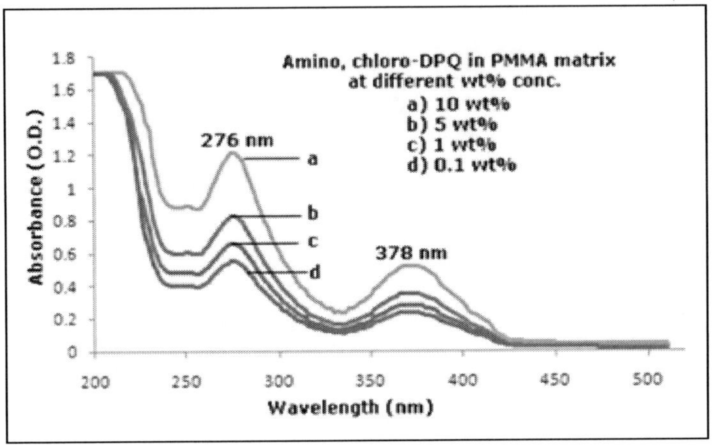

Figure 24: Absorption spectra of amino, chloro-DPQ in PMMA matrix at different wt%.

The peak at lower wavelength is attributed to $\pi \rightarrow \pi^*$ transition while the peak at higher wavelength is attributed to $n \rightarrow \pi^*$ transitions in various molecular orbitals of amino, chloro-DPQ. The amount of absorption depends on the wavelength of the radiation and the structure of the compound. The absorption peak position in the blended films is found to be independent of the dopant concentration. These results indicate that the absorption curve is a characteristic of DPQ. However, absorption intensity of blended films with 10 wt% shows maximum intensity under the chosen experimental conditions, inferring that the doped amino, chloro-DPQ have good compatibility at higher wt%. The absorption peak intensity of amino, chloro-DPQ in PMMA increased in the order of 10% > 5% > 1% ~ 0.1%.

7.2.4. Absorption Spectra of Amino, Chloro-DPQ in PS

The optical absorption spectra of amino, chloro-DPQ in PS film with variation in concentration is shown in Figure 25. amino, chloro-DPQ in PS have two broad absorption bands at 276 nm, attributed to $\pi \rightarrow \pi^*$ transition and 374 nm attributed to $n \rightarrow \pi^*$ transitions. in various molecular orbitals of amino, chloro-DPQ.

The polymeric compound has nearly identical absorption maxima (λ_{max}) at 276 nm with no accountable shift in absorption peak position λ_{max} even with the change in doping concentration.However, a slight change in the intensity of the absorption peak was observed, indicating that the absorption curve is a characteristic of DPQ. The absorption spectra is a superposition of the polymeric compound with polymer matrices. An effect of protonation on optical absorption has previously been observed in derivatives of diphenyl quinoline. The polymeric compound has nearly identical absorption maxima (λ_{max}) in both cases of dichloro-DPQ with PMMA and PS matrices.

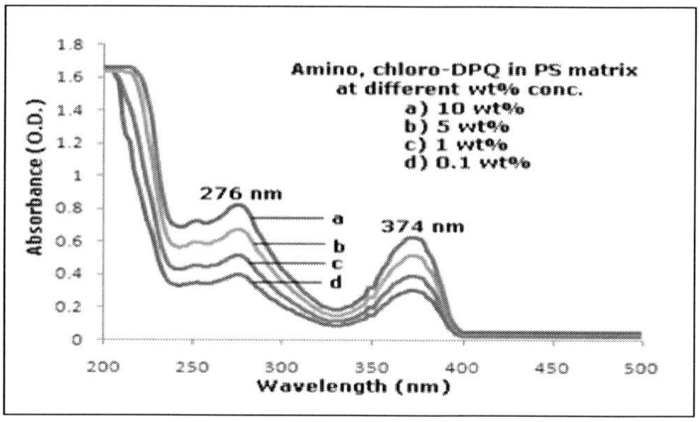

Figure 25. Absorption spectra of amino, chloro-DPQ in PS matrix at different wt%.

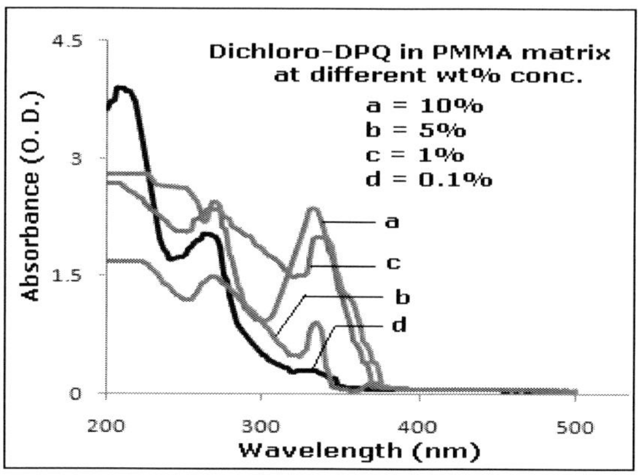

Figure 26. Absorption spectra of dichloro-DPQ with PMMA at different wt%.

7.2.5. Absorption Spectra of Di Chloro-DPQ in PMMA

The optical absorption spectra of dichloro-DPQ in PMMA film with variation in concentration is shown in Figure 26. It displays two broad absorption bands at 276 nm and 340-50 nm.

The former band can be assigned to π- π* transition, which is largely of charge transfer character, transition while the peak at higher wavelength is attributed to n→π* transitions in various molecular orbitals of dichloro-DPQ. The absorption intensity of blended films with 10 wt% shows maximum intensity under the chosen experimental conditions, inferring that the doped dichloro-DPQ have good compatibility at higher wt%.

7.2.6. Absorption Spectra of Dichloro-DPQ in PS

The optical absorption spectra of dichloro-DPQ in PS film with variation in concentration is shown in Figure 27. It displays two broad absorption bands at 276 nm and 330-340 nm. The former band can be assigned to π- π* transition whereas the lowest-energy band peaks at higher wavelength can be attributed to n→π* transitions in various molecular orbitals of dichloro-DPQ with maximum intensity at higher wt%.

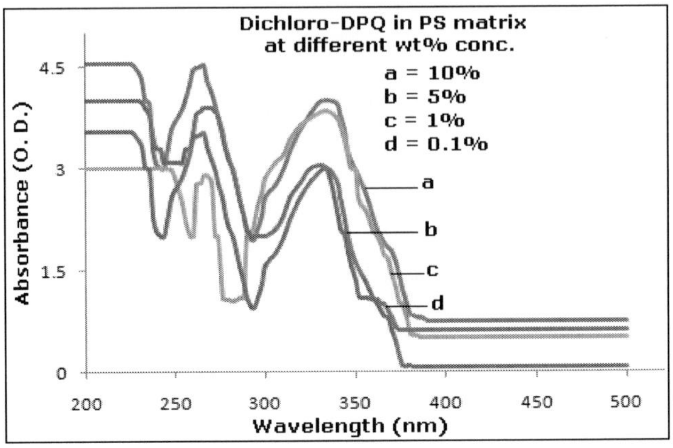

Figure 27. Absorption spectra of dichloro-DPQ with PS at different wt%.

7.2.7. Absorption Spectra of Trichloro-DPQ in PMMA

Figure 28 show the absorption spectra of trichloro-DPQ+PMMA for different concentrations of wt %, displaying two broad absorption bands at 274 nm and 350 nm. Though the spectra of trichloro-DPQ+PMMA at different wt % appear similar, variation in intensity may be due to good compatibility of the complex with the binding at different wt%.

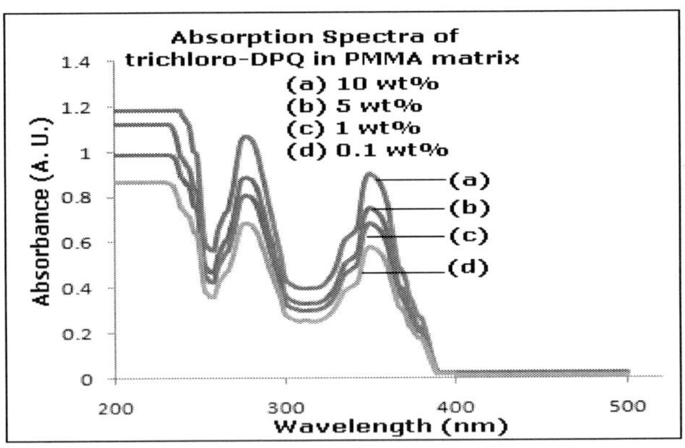

Figure 28. Absorption spectra of trichloro-DPQ with PMMA at different wt%.

7.2.8. Absorption Spectra of Trichloro-DPQ in PS

Figure 29 show the absorption spectra of trichloro-DPQ+PS for different concentrations i.e., 10, 5, 1 and 0.1 wt%. It displays two broad absorption bands at 274 nm and 350 nm.

The former band can be assigned to π- π^* transition of the conjugated polymer main chains, while the peak at higher wavelength is attributed to $n \rightarrow \pi^*$ transitions to the conjugated side chains in various molecular orbitals of dichloro-DPQ.

The absorption peaks in the spectra of different wt % are similar, due to the presence of same chromophore DPQ. Even in this case, absorption intensity of blended films with 10 wt% shows maximum intensity under the chosen experimental conditions, inferring a good deal of interaction between trichloro-DPQ and PS.

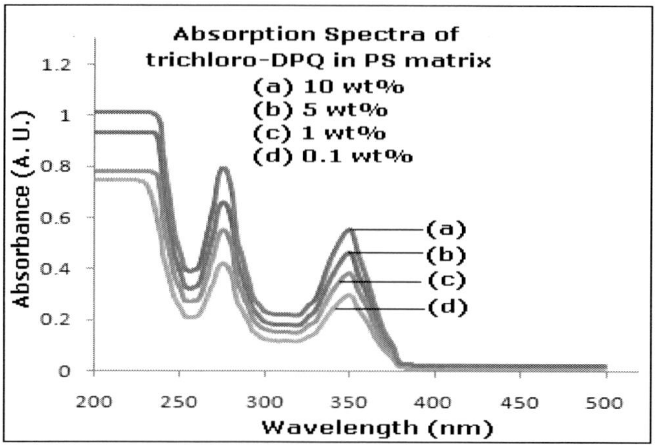

Figure 29. Absorption spectra of trichloro-DPQ with PS at different wt%.

7.3. Photoluminescence Spectra

Luminescence is the cold light emission, which occurs when electron return back from higher energy state to lower energy state at normal or low temperatures. In this phenomenon, the excited energy is always greater than energy of the emitted light. A luminescence spectrum is either band or line spectra or combination of these two spectra.

The luminescence of organic compounds is essentially based on localized π-electron systems within individual organic molecules. In organic luminescent compounds; in contrast, it is the π -electron system of individual molecules that are responsible for luminescence. Therefore, even when organic crystals melt into amorphous aggregates, luminescence still persists. Photoluminescence is the process in which the luminescence is due to photon excitation.

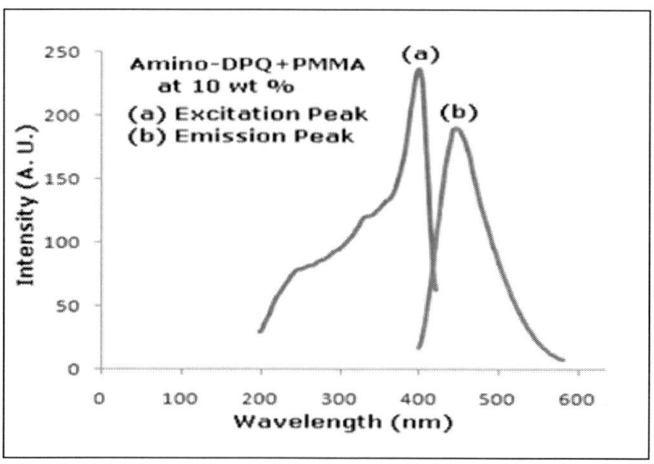

Figure 30. PL spectra of amino-DPQ +PMMA (10%) (Film).

7.3.1. Photoluminescence Spectra of Amino-DPQ in PMMA

The PL spectra of amino-DPQ with PMMA thin films for 10 and 0.1 wt % are investigated. In case of amino-DPQ+PMMA at 10 wt %, the excitation spectrum shows a peak at 385 nm with emission centered at 434 nm as shown in Figure 30.

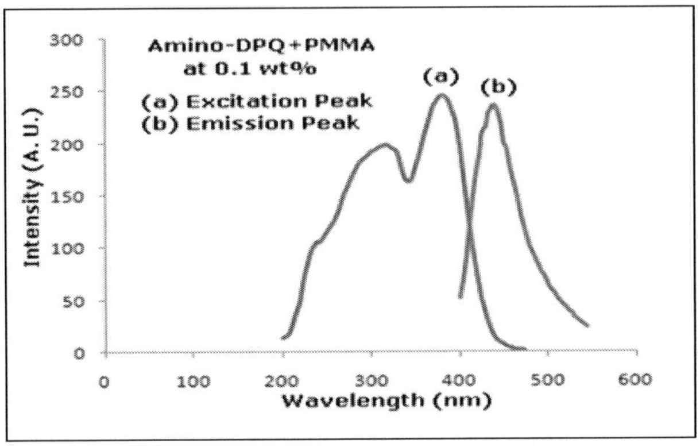

Figure 31. PL spectra of amino-DPQ +PMMA (0.1%) (Film).

However, bathochromic shift by 52 nm is observed in case of amino-DPQ as compared to its PL spectra in powder form[72]. At 0.1 wt % of amino-DPQ+PMMA blended films, the excitation spectrum shows a peak at 385 nm with emission peaking at 434 nm as shown in Figure 31.

The full width half maxima (FWHM) of these blended films was found to be very less, indicating a narrow spectral line emission from the complex.

7.3.2. Photoluminescence Spectra of Amino-DPQ in PS

The polymeric chromophore amino-DPQ+PS at 10 wt % emits intense blue emission of wavelength λ_{max} at 436 nm, when excited at 397 nm as shown in Figure 32. Similarly, the polymeric chromophore amino-DPQ+PS at 0.1 wt % emits intense blue emission of wavelength λ_{max} at 415 nm, when excited at 363 nm as shown in Figure 33.

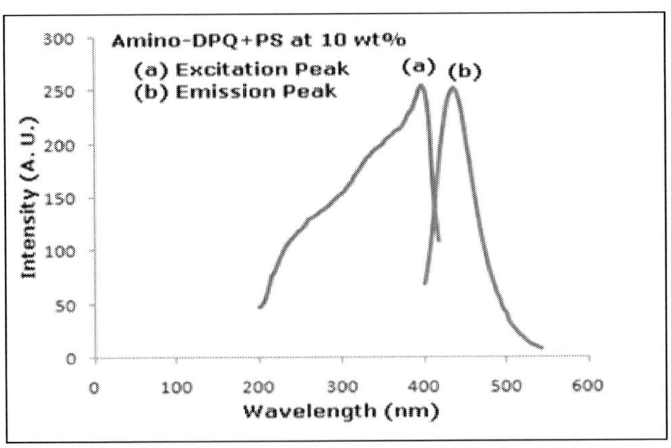

Figure 32. PL spectra of amino-DPQ +PS (10%) (Film).

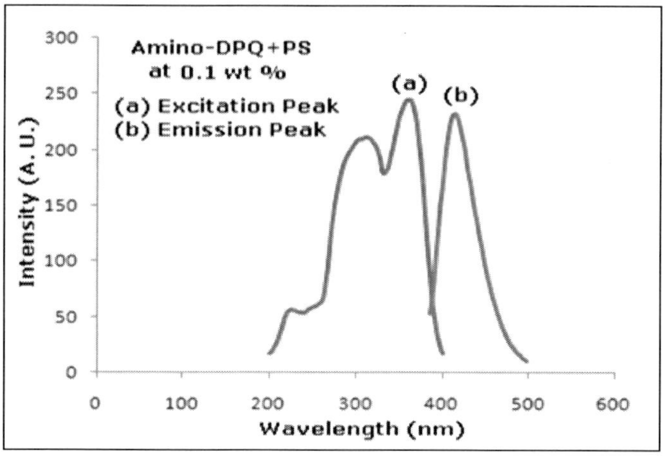

Figure 33. PL spectra of amino-DPQ +PS (0.1%) (Film).

Thus the blended films of amino-DPQ with PS with different wt%, 10 and 0.1%, display peaks at at 436 (λ_{ex} = 397) and 415 (λ_{ex} = 363), respectively with maximum intensity at 0.1%. When the emission spectra of amino-DPQ powder are compared with blended thin films of PMMA and PS matrixes of different wt%, hypsochromic shift was observed [71]. These blended films can be used as blue light emitting phosphors for blue OLEDs and flat panel displays.

7.3.3. Photoluminescence Spectra of Amino Chloro-DPQ in PMMA

The excitation spectra at different concentrations such as 10 and 0.1 wt% in PMMA exhibit same nature of curve having excitation peaks at 380 nm for dopant with PMMA.

In case of blended thin films of amino, chloro-DPQ in PMMA matrix, PL emission peaks centred at 458 nm and 450 nm at different weight % concentration such as 10, and 0.1 wt%, respectively as shown in Figure 34 and 35.

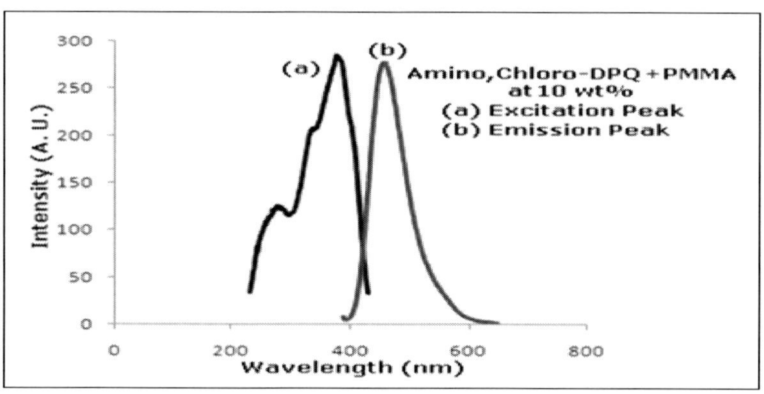

Figure 34. PL spectra of amino, chloro-DPQ +PMMA (10%) (Film).

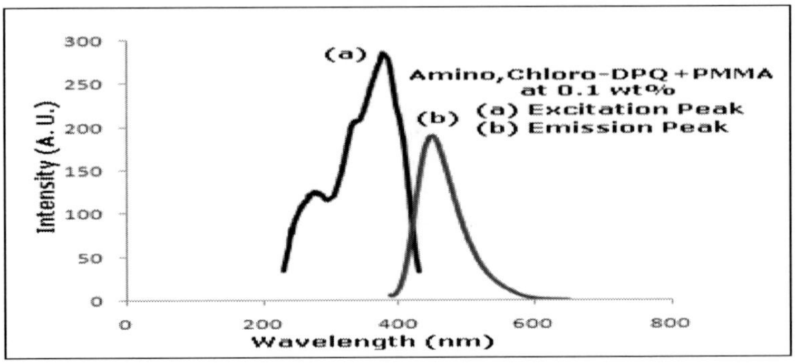

Figure 35. PL spectra of amino, chloro-DPQ+PMMA (0.1%) (film).

7.3.4. Photoluminescence Spectra of Amino Chloro-DPQ in PS

The excitation spectra (inset) of amino, chloro-DPQ+PS at 10 and 0.1 wt % are shown in Figure 36 and 37, respectively. When these films are excited at 380 nm, emission peaks were observed at 444nm and 434 nm, respectively. This shows hypsochromic shift in case of both amino, chloro-DPQ in host polymer matrices (PMMA and PS) as compared to the amino, chloro-DPQ in powder form. A light emitting polymer blended with a photonically inert polymer such as PMMA and PS often yields a higher PL intensity on photoexcitation due to the dilution effect.

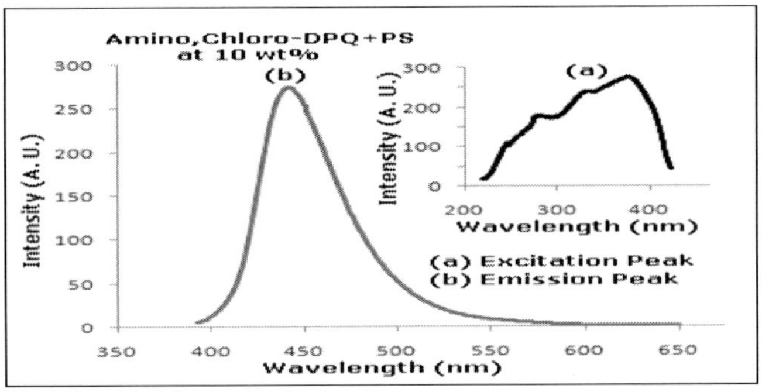

Figure 36. PL spectra of amino, chloro-DPQ+PS (10%) (film).

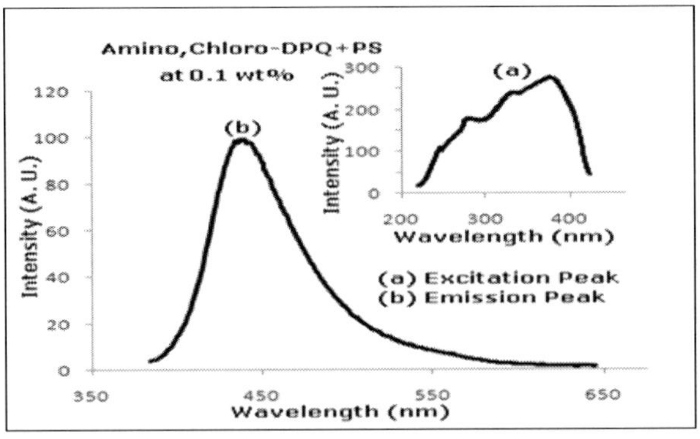

Figure 37. PL spectra of amino, chloro-DPQ+PS (0.1%) (film).

7.3.5. Photoluminescence Spectra of Dichloro-DPQ in PMMA

The PL spectra of thin films of the dichloro-DPQ with PMMA are investigated for 10 and 0.1 wt% are shown in Figures 38 and 39, respectively. In case of polymeric chromophore dichloro-DPQ+PMMA at 10 weight %, the excitation spectrum shows a peak at 365 nm with blue light emission at 443 nm. At 0.1 weight % with combination of dichloro-DPQ with PMMA, the polymeric chromophore emits intense blue emission wavelength of 421 nm, when excited at 365 nm.

When the PL spectra of dichloro-DPQ with PMMA is compared for different weight concentrations at 10 and 0.1%, it displays emission peak at 443 and 421 nm, respectively with maximum intensity for 0.1 wt% under the same excitation at 365 nm for all concentrations.

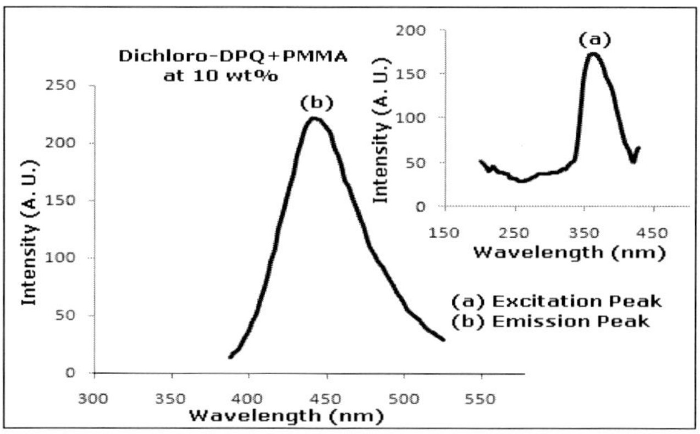

Figure 38. PL spectra of dichloro-DPQ+PMMA (10%) (Film).

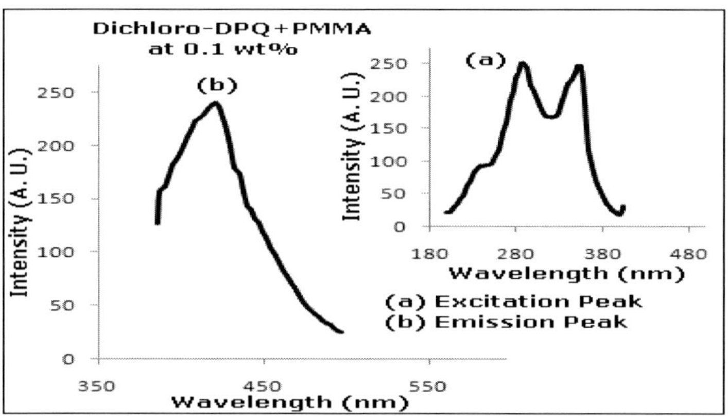

Figure 39. PL spectra of dichloro-DPQ+PMMA (0.1%) (Film).

7.3.6. Photoluminescence Spectra of Dichloro-DPQ in PS

PL spectra of dichloro-DPQ+PS at different wt % such as 10 and 0.1 wt % are shown in Figure 40 and 41. In case of dichloro-DPQ+PS at 10 wt%, the polymeric chromophore emits intense blue emission of wavelength 422 nm, when excited at 355 nm. At 0.1 wt %, the excitation spectrum shows a peak at 355 nm with emission centered at 393 nm. A small hypsochromic shift is observed in case of blended thin films of dichloro-DPQ+PMMA and dichloro-DPQ+PS when compared to the dichloro-DPQ powder, which may be due to molecular interactions between the complexes and the binding polymer. However, both the PMMA and PS matrices with dichloro-DPQ emit intense blue emission, depending on the concentrations of the materials.

The blue emission band is at the different locations as the single chain emission of dichloro-DPQ in a matrix of PMMA and PS. However, the single – chain emission of even excimer-forming conjugated polymers can be usually observed in dilute fluid or solid solutions.

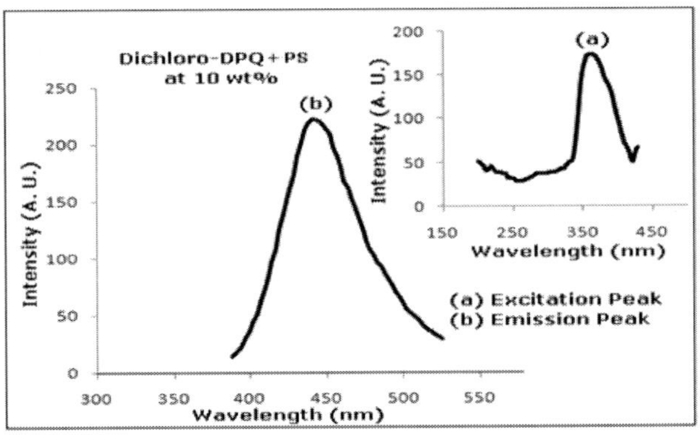

Figure 40. PL spectra of dichloro-DPQ+PMMA (10%) (Film).

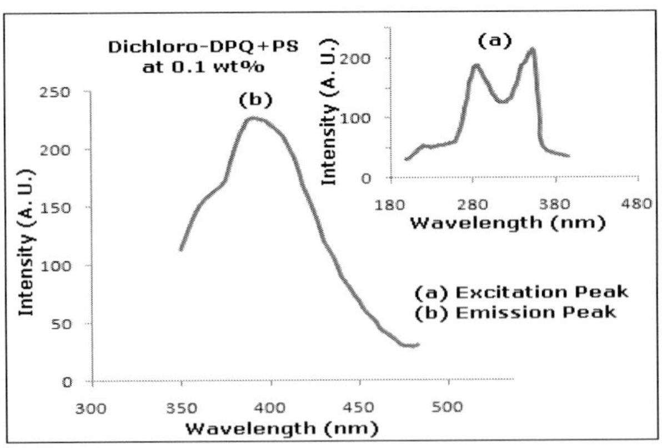

Figure 41. PL spectra of dichloro-Dpq+Ps (0.1%) (Film).

7.3.7. Photoluminescence Spectra Of trichloro-DPQ in PMMA

PL spectra of trichloro-DPQ+PMMA at different wt % are shown in Figure 42 and 43, respectively.

The excitation spectra at different concentrations such as 10 and 0.1 wt% in PMMA matrix show same nature of curve with excitation peak at 350 nm.

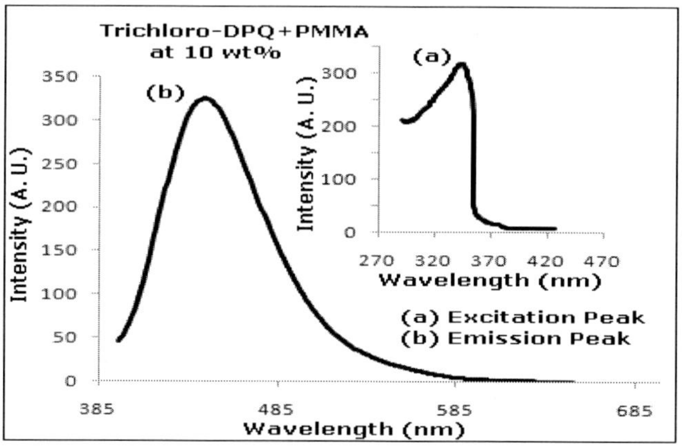

Figure 42. PL spectra of trichloro-DPQ+PMMA at 10% (film).

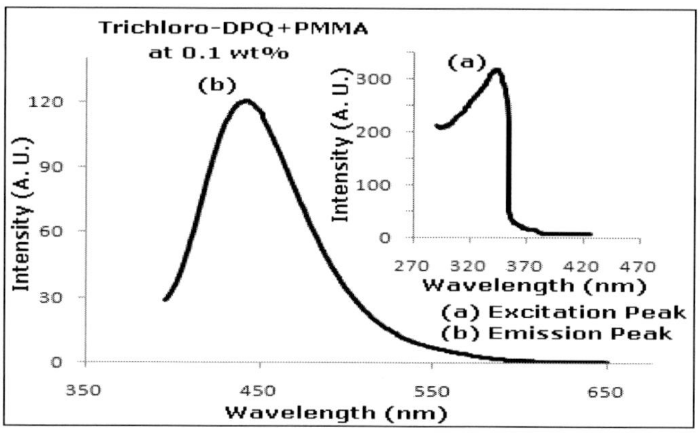

Figure 43. PL spectra of trichloro-DPQ+PMMA at 0.1% (film).

7.3.8. Photoluminescence Spectra of Trichloro-DPQ in PS

PL spectra ofn trichloro-DPQ with PS at different wt % are shown in Figure 44 and 45, respectively. In both the cases, emission peak was observed at 429 nm when excited at 345 nm with blue shift at 31 nm as compared to the powder form of trichloro-DPQ. It is generally true that the emission spectrum has much better resolved structure than is seen in the absorption spectrum.

However, the spectrum is significantly broad, which may be due to additional π–π* interactions or conformational rigidity in the solid state. Hypsochromic shift is obtained in the emission with decreasing concentration which may be due to molecular interactions. This shift in PL may be due to lesser dipole moment of the compound in the excited state.

The PL is originated from the radiative recombination of singlet exciton, especially singlet polaron–exciton i.e., bound electron–hole pairs in polymers. Spectral data of optical absorbance and emission spectra of DPQ derivatives are tabulated in Table 4.

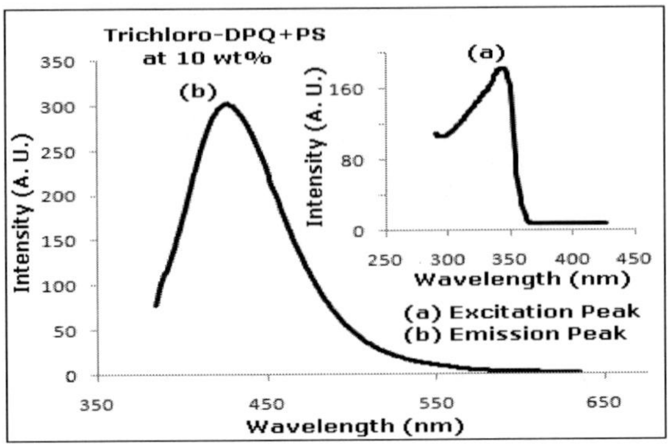

Figure 44. PL spectra of trichloro-DPQ+PS at 10% (film).

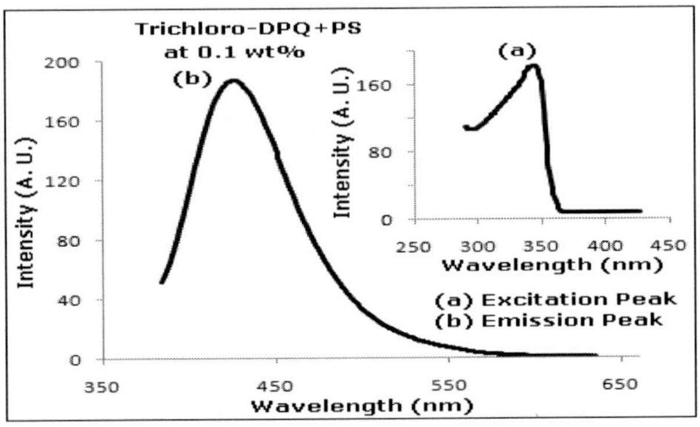

Figure 45. PL spectra of trichloro-DPQ+PS at 0.1% (film).

Table 4. Spectral data of optical absorbance and emission spectra of DPQ derivatives

Complex	Concentration (%)	Absorption Peaks in PMMA (nm)		Absorption Peaks in PS(nm)		λ_{exc} in PMMA (nm)	λ_{emi} in PMMA (nm)	λ_{exc} in PS (nm)	λ_{emi} in PS (nm)
Amino DPQ	10	I	268	I	268	400	450	397	436
		II	360	II	390				
	5	I	268	I	268	397	445	386	426
		II	350	II	380				
	1	I	268	I	268	385	440	375	423
		II	340	II	385				

Complex	Concentration (%)	Absorption Peaks in PMMA (nm)		Absorption Peaks in PS(nm)		λ_{exc} in PMMA (nm)	λ_{emi} in PMMA (nm)	λ_{exc} in PS (nm)	λ_{emi} in PS (nm)
	0.1	I	268	I	268	382	434	363	415
		II	330	II	380				
Amino-chloro DPQ	10	I	276	I	276	380	458	376	444
		II	378	II	374				
	5	I	276	I	276	380	457	376	440
		II	378	II	374				
	1	I	276	I	276	380	450	376	435
		II	378	II	374				
	0.1	I	276	I	276	380	450	376	435
		II	378	II	374				
Dichloro DPQ	10	I	260	I	276	365	443	355	422
		II	334	II	330				
	5	I	260	I	276	365	438	355	434
		II	334	II	330				
	1	I	260	I	276	365	435	355	426
		II	334	II	330				
	0.1	I	260	I	276	365	421	355	393
		II	334	II	330				
Trichloro DPQ	10	I	274	I	274	350	444	350	429
		II	350	II	350				
	5	I	274	I	274	350	444	350	429

Table 4. (Continued)

Complex	Concentration (%)	Absorption Peaks in PMMA (nm)		Absorption Peaks in PS(nm)		λ_{exc} in PMMA (nm)	λ_{emi} in PMMA (nm)	λ_{exc} in PS (nm)	λ_{emi} in PS (nm)
		II	350	II	350				
	1	I	274	I	274	350	444	350	429
		II	350	II	350				
	0.1	I	274	I	274	350	444	350	429
		II	350	II	350				

8. Key Challenges

With all of their advantages, OLEDs are still more expensive than the existing displays, having operational lifetime challenges and colour balance issues, particularly in blue. Blue OLED compounds are the most susceptible to aging and have a 50% drop in brightness in about 14,000 hours. The drop in brightness of blue OLEDs happens much faster than red and green OLEDs which cause colour balance issues in OLEDs [137-138]. There are certain key challenges ahead in order to commercialize OLEDs for large screen applications, which include- electricity to light conversion efficiency, device stability and life time (especially blue colour), material selection and optimization, encapsulation manufacturing cost, fine patterning, contrast, pixel switching, colour saturation. OLED seems to be the perfect technology for all types of displays, but still face the following hurdles:

8.1. Driving Voltage

The minimum voltage require to operate an OLED is known as driving voltage. The driving voltage of a device can be minimized by employing low work function cathode, a high quality hole and electron injection layer, proper device configuration, energy efficient emissive materials and by inserting an effective buffer layer between the organic layer and anode.

8.2. Lifetime

Red and green OLED films have longer lifetimes (46,000 to 230,000 hours), blue organics currently have much shorter lifetimes (up to around 14,000 hours).

8.3. Encapsulation

Non-emissive dark spots develop initially and lead to the degradation of the device. This limits the lifetime and hence the device needs protection by a hermetic encapsulation - the process of bonding a metal sheet on to the substrate glass using UV cured epoxy. It is carried out in a glove box, which is free from oxygen and water.

8.4. Degradation Issues

Degradation in OLED has been attributed to various mechanisms such as electro chemical reactions at the electrode/organic interface, crystallization of organic solids, migration of ionic species [139-143]. Degradation due to the growth of dark spots can be attributed to the dust particles deposited during the fabrication process, diffusion of oxygen leads to oxidation thereby causing delamination of electrodes, impurities that exist in the organic materials, decomposition of light emitting materials. Water can easily damage OLEDs because they are extremely sensitive to moisture

8.5. Improvement in Quantum Efficiency

Quantum efficiency is defined as the ratio of number of photons emitted to the number of photons injected. Quantum efficiency depends on the photon to electron ratio. Hence materials with good quantum efficiency are to be selected during OLED manufacture.

8.6. Manufacturing

Manufacturing processes are expensive right now.

Conclusion

Solid state lighting has proved to be vital important for existing and future challenges such as sustainable development, energy and community health as well as for improving the quality and standard of life. Hence, the interest in solid state light sources for general illumination has been rapidly growing, boosted by the search for cheaper light sources and the emergent concern on global energy consumption. But, accomplishment of pure blue light with efficiency, colour purity and stability remains a challenge. Review on OLEDs and its materials reveals that electricity to light conversion efficiency, device stability and life time, especially blue colour, material selection and optimization, encapsulation to prevent device from water, oxygen and any reacting chemical, uniformity over large areas, manufacture cost, especially for large area, fine patterning, contrast, pixel switching, colour saturation are the challenges presently ahead in order to compete with the present lighting system with the eco-friendly OLEDs. Significant efforts have been made at global level on development and fabrication of red, green and blue emitting complexes and devices. However, further improvements in efficiency are required to take over a large proportion of the flat panel display market. After peer review of literature, we have reported the synthesis and characterization of DPQ derivatives, which emit intense blue light emitting materials. The conclusions from the above studies are as follows:

The low cost synthesis and characterization of amino, amino-chloro, dichloro and trichloro substituted derivatives of diphenyl quinoline (DPQ) by Friedlander condensation and the preparation of blended thin films with the synthesized complexes and binding polymers such as poly (methyl methacrylate) (PMMA) and polystyrene (PS) at different weight % concentrations yielded good phosphors and blended films with emission wavelength centered in the blue region. FTIR analysis confirms the formation and molecular structure of the synthesized DPQ-derivatives. X-ray diffractograms reveals that the synthesized organic compounds are crystalline in nature. These outstanding properties make amino-DPQ, attractive candidates as advanced materials in the field of photonics and electronics. Thermal properties of the materials are analyzed by TGA and DTA results conclude that the starting materials are affordable for the chemical synthesis of DPQ derivatives; they are thermally stable. Among all the synthesized complexes amino-chloro DPQ exhibits maximum thermal stability upto 540.16^{0}C. However, the the other derivatives of DPQ ate thermally stable in the temperature range of 299-395^{0}C. Optical properties of these complexes projects blue light emission in blue region of EM spectrum. Among all the DPQ complexes trichloro DPQ displays maximum intensity with emission peak at 460 nm. Hence, these complexes can be used as emissive material for the fabrication of OLEDs by vacuum evaporation technique.

In order to study the compatibility of the synthesized DPQ derivatives, they are molecularly doped in binding polymers PMMA and PS. The resultant films were found to exhibit excellent transparency with emission peak in the range of 421-458 A^0 in PMMA and 393-444 A^0 in PS. Among all the blended films aminochloro DPQ emits maximum intensity in PMMA as well as PS for 10 wt% with full width half maxima (FWHM) less than 10 nm , indicating a narrow spectral line emission from the films. Hence, these materials can be considered as an ultimate substitute for the existing blue organic light-emitting materials, generally employed for fabricating OLEDs, display devices, solid state lighting and many more applications by vacuum evaporation as well as by solution techniques. In near future the above said challenges will be surmounted by the researchers and will generate a new market in display technology. Hence there is a need of continuous efforts to explore new blue emitting materials based on the quinoline backbone.

REFERENCES

[1] J. Kido, M. Kimura, K. Nagai, *Science,* 267 (1995) 1332.
[2] S. Reineke, F. Lindner, G. Schwartz, N. Seidler, K. Leo, *Nature,* 459 (2009) 234.
[3] R.S. Sun, S.R. Forrest, *Appl. Phys. Lett.,* 91 (2007) 263503.
[4] Y.R. Sun, N.C. Giebink, H. Kanno, B.W. Ma, M.E. Thompson, S.R. Forrest, *Nature,* 440 (2006) 908.

[5] M.A. Baldo, D.F. O'Brien, Y. You, A. Shoustikov, S. Sibely, M.E. Thompson, S.R. Forrest, *Nature,* 395 (1998) 151.
[6] D.Y. Kim, H.N. Cho, C.Y. Kim, *Prog. Polym. Sci.,* 25 (2000) 1089– 1139.
[7] T.C. Tsai, W.Y. Hung, L.C. Chi, K.T. Wong, C.C. Hsieh, P.T. Chou, *Org. Electron.,* 10 (2009) 158–162.
[8] Q.J. Qi, X.M. Wu, Y.L. Hua, Q.C. Hou, M.S. Dong, Z.Y. Mao, B. Yin, S.G. Yin, *Org. Electron.,* 11 (2010) 503–507.
[9] Huaijun Tang, Yanhu Li, Caihong Wei, Bing Chen, Wei Yang, Hongbin Wu, *Yong CaoDyes and Pigments,* 91 (2011) 413-421.
[10] C. Hosokawa, H. Higashi, H. Nakamura, T. Kusumoto, *Appl. Phys. Lett.,* 67 (1995) 3853.
[11] Y. Liu, J.H. Guo, J. Feng, H.D. Zhang, Y.Q. Li, Y. Wang, *Appl. Phys. Lett.,* 78 (2001) 2300.
[12] W.B. Im, H.K. Hwang, J.G. Lee, K. Han, Y. Kim, *Appl. Phys. Lett.,* 79 (2001) 1387.
[13] H.T. Shih, C.H. Lin, H.H. Shih, C.H. Cheng, *Adv. Mater.,* 14 (2002) 1409.
[14] C.C. Wu, Y.T. Lin, H.H. Chiang, T.Y. Cho, C.W. Chen, K.T.Wong, Y.L. Liao, G.H. Lee, S.M. Peng, *Appl. Phys. Lett.,* 81 (2002) 577.
[15] Y.H. Kim, D.C. Shin, S.H. Kim, C.H. Ko, H.S. Yu, Y.S. Chae, S.K. Kwon, *Adv. Mater.,* 13 (2001) 1690.
[16] J.M. Shi, C.W. Tang, *Appl. Phys. Lett.,* 80 (2002) 3201.
[17] Y. Kan, L.D. Wang, L. Duan, Y.C. Hu, G.S. Wu, Y. Qiu, *Appl. Phys. Lett.,* 84 (2004) 1513.
[18] Jun Xie, Chunyan Chen, Shufen Chen, Yang Yang, Ming Shao, Xu Guo,
[19] Quli Fan , Wei Huang; *Organic Electronics,* 12 (2011) 322–328.
[20] M. Nohara, M. Hasegawa, C. Hosokawa, H. Tokailin, T. Kusumoto, *Chem. Lett.,* (1990) 189.
[21] C. Adachi, T. Tsutsui, S. Saito, *Appl. Phys. Lett.,* 56 (1990) 799.
[22] M. Uchida, Y. Ohmori, C. Morishima, K. Yoshino, *Synth. Met.,* 55 (1993) 4168.
[23] D.D. Gebler, Y.Z. Wang, J.W. Blstchford, S.W. Jessen, L.B. Lin, T.L. Gustafson, H.L. Wang, T.M. Awagwr, A.G. MacDiarmid, A.J. Epstein, *J. Appl. Phys.,* 78 (1995) 4264.
[24] G. Grem, G. Leditzky, B. Ullrich, G. Leising, *Adv. Mater.,* 4 (1992) 36.
[25] J. Kito, K. Hongawa, K. Okuyama, K. Nagai, *Appl. Phys. Lett.,* 63 (1993) 2627.
[26] W. Tachelet, S. Jacobs, H. Ndayikengurukiye, H.J. Geise, *Appl. Phys. Lett.,* 64 (1994) 2364.
[27] Edwards, S. Blumstengel, I. Sokolik, R. Dorsinville, H. Yun, T.K. Kwei, Y. Okamoto, *Appl. Phys. Lett.,* 70 (1997) 298.
[28] Y. Hamada, T. Sato, M. Fujita, T. Fujii, Y. Nishio, K. Shibata, *Jpn. J. Appl. Phys.,* 32 (1993) 511.
[29] Gui Yu, Yunqi Liu, Yaru Song, Xia Wu, Daoben Zhu; *Synthetic Metals,* 117 (2001) 211-214
[30] Silu Tao, Yilan Jiang, Shiu-Lun Lai, Man-Keung Fung, Yechun Zhou, Xiaohong Zhang c, Weiming Zhao , Chun-Sing Lee, *Organic Electronics,* 12 (2011) 358–363.
[31] D. H. Dunlap, P. E. Parris, and V. M. Kenkre, *Phys. Rev. Lett.,* 77 (1996) 542.
[32] G. horowitz, D. Fichou, X. Z. Peng, Z. Xu, and F. Garnier, *Solid State Commun.,* 72 (1989) 381.
[33] R. H. Friend, J. Burroughes, T. Shimoda, Polymer diodes, *Phys. World,* 12 (1999) 35.

[34] R. McNeill, R. Siudak, J. H. Wardlaw, D. E. Weiss, *Australian Journal of Chemistry,* 16 (1963) 1056.

[35] Bernanose, M. Comte, P. Vouaux, *J. Chim. Phys.,* 50 (1953) 64.

[36] Bernanose, P. Vouaux, *J. Chim. Phys.,* 50 (1953) 261.

[37] Bernanose, *J. Chim. Phys.,* 52 (1955) 396.

[38] Bernanose, P. Vouaux, *J. Chim. Phys.,* 52 (1955) 509.

[39] E. Gurnee, R. Fernandez, *US Patent,* 3 172 862, 1965.

[40] M. Pope, H. Kallman, P. Magnante, *J. Chem. Phys.,* 38 (1963) 2042.

[41] W. Digby, M. Schadt, *US Patent,* 3 621 321, 1971.

[42] R. Partridge, *US Patent,* 3 995 299, 1976.

[43] P. Vincett, W. Barlow, R. Hann, G. Roberts, *Thin Solid Films,* 94 (1982) 171.

[44] Tang, *US Patent,* 4 164 431, 1979.

[45] Tang, *US Patent,* 4 356 429, 1982.

[46] Tang, S. VanSlyke, *Appl. Phys. Lett.,* 51 (1987) 913.

[47] J. Burroughes, D. Bradley, A. Brown, R. Marks, K. Mackay, R. Friend, P. Burn, A. Holmes, *Nature,* 347 (1990) 539.

[48] R. Wessling, R. Zimmerman, *US Patent,* 3 401 152, 1968.

[49] R. Wessling, R. Zimmerman, *US Patent,* 3 706 677, 1972.

[50] R. Wessling, *J. Polym. Sci. Polym. Symp.,* 72 (1985) 55.

[51] R. Friend, J. Burroughes, D. Bradley, WO Patent 90/13 148, 1990. R. Friend, J. Burroughes, D. Bradley, *US Patent,* 5 247 190, 1993.

[52] Braun, A. Heeger, *Appl. Phys. Lett.,* 58 (1991) 1982.

[53] H. Radousky, A. Madden, K. Pakbaz, T. Hagler, H. Lee, H. Lorenzana, G. Fox, P. Elliker, *Int. SAMPE Tech. Conf.,* 27 (1995) 1143.

[54] J. Scott, J. Kaufman, P. Brock, R. DiPietro, J. Salem, J. Goitia, *J. Appl. Phys.,* 79 (1996) 2745.

[55] Y. Ohmori, M. Uchida, K. Muro, K. Yoshino, *Jpn. J. Appl. Phys.,* 30 (1991) L1938.

[56] Y. Ohmori, C. Morishita, M. Uchida, K. Yoshino, *Jpn. J. Appl. Phys.,* 31 (1992) L568.

[57] D. Braun, G. Gustaffson, D. MacBranch, A. Heeger, *J. Appl. Phys.,* 72 (1992) 564.

[58] Grem, G. Leditzky, B. Ullrich, G. Leising, *Adv. Mater.,* 4 (1992) 36.

[59] Y. Ohmori, M. Uchida, K. Muro, K. Yoshino, *Jpn. J. Appl. Phys.,* 30 (1991) L1941.

[60] M. Fukuda, K. Sawada, S. Morita, K. Yoshino, *Synth. Met.,* 41 (1991) 855.

[61] Hosokawa, N. Kawasaki, S. Sakamoto, T. Kusumoto, *Appl. Phys. Lett.,* 61 (1992) 2503.

[62] J. Kido and M. Kohda, K. Okuyama, K. Nagai, *Appl. Phys. Lett.,* 61 (1992) 761.

[63] K. Agrawal, S. A. *Jenekhe Macromolecules,* 26 (1993) 895.

[64] Z. Yang, I. Sokolik, F. E. *Karasz, Macromolecules,* 26 (1993) 1188.

[65] J. A. Osaheni and S. A. Jenekhe, *Macromolecules,* 27 (1994) 739.

[66] J. Kido, K. Hongawa, K. Okuyama, K. Nagai, *Appl. Phys. Lett.,* 64 (1994) 815.

[67] Guangming Wang, Chunwei Yuan, Haiming Wu, Yu Wei *J. Appl. Phys.,* 78 (1995) 2679.

[68] M. Aguiar, F . E. Karasz, L. Akcelrud, *Macromolecules,* 28 (1995) 4598.

[69] Yang Yang and Qibing Pei, *Appl. Phys. Lett.,* 68 (19960 2708.

[70] P. W. M. Blom, M. J. M. de Jong, J. J. M. Vleggaar, *Appl. Phys. Lett.,* 68 (1996) 3308.

[71] Schulz, Y. Kaminorz, L. Brehmer, *Synthetic Metals,* 84 (1997) 449.

[72] Rost, A. Teuschel, S. Pfeiffer, H.-H. Horhold, *Synthetic metals,* 84 (1997) 269.

[73] F. Cacialli, C. -M. Bouch, P. Le Barny, R.H. Friend, H. Facoetti, F. Soyer, P. Robin, *Optical Materials,* 9 (1998) 163.
[74] S. E. Shaheen, G. E. Jabbour, M. M. Morrell, Y. Kawabe, B. Kippelen, N. Peyghambarian, *J. Appl. Phys.,* 84 (1998) 2324.
[75] S. Shetty, E. B. Liu, R. J. Lachicotte, and S. A. Jenekhe, *Chem. Mater.,* 11 (1999) 2292.
[76] X. Zhang, A. S. Shetty, and S. A. Jenekhe, *Macromolecules,* 32 (1999) 7422.
[77] Sin-Hye Jung, Jae-Ho Choi, Soon-Ki Kwon, Won-Jei Cho, Chang-Sik Ha, *Thin Solid Films,* 363 (2000) 160.
[78] Z. He, G. H. W. Milburn, K. J. Baldwin, D.A. Smith, A. Danel, P. Tomasik, *Journal of Luminescence,* 86 (2000) 1.
[79] Min Ho Kim, Jung-Il Jin, Chul Joo Lee, Nakjoong Kim, and Ki Hong Park, *Bull. Korean Chem. Soc.,* 23 (2002) 964.
[80] X. Zhan, Y. Liu, X. Wu, S. Wang, D. Zhu, *Macromolecules,* 35 (2002) 2529.
[81] F. Liang, J. Chen, Y. Cheng, L. Wang, D. Ma, X. Jing, F. Wang, *J. Mater. Chem.,* 13 (2003) 1392.
[82] T. Yamamoto, Y. Saitoh, K. Anzai, H. Fukumoto, T. Yasuda, Y. Fujiwara, B. Choi, K. Kubota, T. Miyamae, *Macromolecules,* 36 (2003) 6722.
[83] J. Ding, J. Gao, Qi Fu, Y. Cheng, D. Ma, L. Wang, Synthetic
[84] *Metals,* 155 (2005) 539.
[85] Fang-Iy Wu, Huei-Jen Su, Ching-Fong Shu, Liyang Luo, Wei-Guang Diau, Chien-Hong Cheng, Jiun-Pey Duan and Gene-Hsiang Lee, *J. Mater. Chem.,* 15 (2005) 1035.
[86] S. W. Tsang, S. K. So, J. B. Xu, *J. Appl. Phys.,* 99 (2006) 013706.
[87] Gondek, I.V. Kityk, A. Danel, A. Wisla, J. Sanetra, *Synthetic Metals,* 156 (2006) 1348.
[88] Lei Zhao, Sheng Biao Li, Gui An Wen, Bo Peng, Wei Huang,
[89] *Materials Chemistry and Physics,* 100 (2006) 460.
[90] S. Calus, E. Gondek, A. Danel, B. Jarosz, J. Niziol, A.V. Kityk, *Materials Science and Engineering B,* 137 (2007) 255.
[91] B.M. Bahirwar, R.G. Atram, R.B. Pode, S.V. Moharil, *Materials Chemistry and Physics,* 106 (2007) 364.
[92] Xiao-Ming Yu, Gui-Jiang Zhou, Ching-Shan Lam, Wai-Yeung Wong, Xiu-Ling Zhu, Jia-Xin Sun, Man Wong, Hoi-Sing Kwok, *Journal of Organometallic Chemistry,* 693 (2008) 1518.
[93] T. W. Kwon, A. P. Kulkarni, S. A. Jenekhe, *Synthetic Metals,* 158 (2008) 292.
[94] Khizar-ul Haq, Liu Shan-peng, M.A. Khan, X.Y. Jiang, Z.L. Zhang, Jin Cao, W.Q. Zhu, *Current Applied Physics,* 9 (2009) 257.
[95] R. Simas, T. D. Martins, T. D .Z. Atvars, L. Akcelrud, *Journal of Luminescence,* 129 (2009) 119.
[96] K. Balamurugan, V. Jeyachandran, S. Perumal, T. H. Manjashetty, P. Yogeeswari, D. Sriramb, *European Journal of Medicinal Chemistry,* 45 (2010) 682.
[97] Wettach, S. S. Jester, A. Colsmann, U. Lemmer, N. Rehmann, K. Meerholz, S. Höger, *Synthetic Metals,* 160 (2010) 691.
[98] V. D. Gupta, V. S. Padalkar, K. R. Phatangare, V. S. Patil, P. G. Umape, N. Sekar, *Dyes and Pigments,* 88 (2011) 378.
[99] S. B. Raut, S. J. Dhoble, R. G. Atram, *Synthetic Metals,* 161 (2011) 391.
[100] S. Tao, Y. Jiang, S. –L. Lai, M. –K. Fung, Y. Zhou, X. Zhang, W. Zhao, C. –S. Lee, *Organic Electronics,* 12 (2011) 358.

[101] de Almeida, Anibal,: Joint Research Centre of the European Commission; 2012. 978-92-79-27365-0.

[102] S.C. Dong, C.H. Gao, X.D. Yuan, L.S. Cui, Z.Q. Jiang, S.T. Lee, L.S. Liao, *Org. Electron.,* 14 (2013) 902.

[103] S. Liu, J. Yu, Z. Ma, J. Zhao, *J. Lumin.,* 134 (2013) 665.

[104] J.R. Koo, S.J. Lee, G.W. Hyung, B.Y. Kim, D.H. Lee,W.Y. Kim, K.H. Lee, S.S. Yoon, Y.K. Kim, *Thin Solid Films,* 544 (2013) 234.

[105] Jian Wang, Fujun Zhang, Jian Zhang, Weihua Tang, Aiwei Tang, Hongshang Peng, Zheng Xu, Feng Teng, Yongsheng Wang, Journal of Photochemistry and Photobiology C: Photochemistry Reviews 17 (2013) 69-104.

[106] N.Thejo Kalyani, S.J.Dhoble; *Ren. Sus. Ene. Rev.,* 32(2014)448–467.

[107] Aníbal De Almeida, Bruno Santos, Bertoldi Paolo, Michel Quicheron *Ren. Sus. Ene. Rev.,* 34 (2014) 30–48.

[108] Joachim Schleich, Bradford Mills, *Elisabeth Dütschke Energy Policy,* 72 (2014) 35–42.

[109] N. Thejo Kalyani, S. J. Dhoble; *Ren. Sus. Ene. Rev.,* 44(2015) 319–347.

[110] Steve Sorrell *Ren. Sus. Ene. Rev.,* 47(2015)74–82

[111] X. Wang, M. Shao, L. Liu, *Synthetic Metals,* 160 (2010) 718.

[112] B.W. D'Andrade, J. Brooks, V. Adamovich, M.E. Thompson, S.R. Forrest, *Adv. Mater.,* 14 (2002) 1032.

[113] S. O. Norris, J. K. Stille, *Macromolecules,* 9 (1976) 496.

[114] Baldacchini, P.Chiacchiaretta, R. Reisfeld, E. Zigansky, *J. of Lumin.,* 129 (2009) 1849.

[115] Zhang, D. Braun, A. J. Heeger, *J. Appl. Phys.,* 73 (1993) 5177.

[116] T. Eicher, S. Hauptmann, The Chemistry of Heterocycles: Structure, Reaction, Syntheses and Applications, 2nd Edition; John Wiley & Sons, Inc.: Chichester, U.K., 2003.

[117] P. Kushto, Y. Iizumi, J. Kido, Z. H. Kafafi, *J. Phys. Chem. A,* 104 (2000) 3670.

[118] X. -M. Wu, Y. -L. Hua, Z. -Q. Wang, J. -J. Zheng, X. -L. Feng, Y. –Y Sun, *Chin. Phys. Lett.,* 22 (2005) 1797.

[119] P. Michael, *Nat. Prod. Rep.,* 14 (1997) 605.

[120] G. Markees, V. C. Dewey, G. W. Kidder, J. Med. Chem. 13 (1970) 324.

[121] R. Abegg, *Amidochrysene. Ber.,* 23 (1890) 792.

[122] T. W. Kwon, M. M. Alam, S. A. Jenekhe, *Chem. Mater.,* 16 (2004) 4657.

[123] S. Shetty, E. B. Liu, R. J. Lachicotte, S. A. Jenekhe, *Chem. Mater.,* 11 (1999) 2292.

[124] S. A. Jenekhe, X. Zhang, X. L. Chen, V.-E. Choong, Y. Gao, B. R. Hsieh, *Chem. Mater.,* 9 (1997) 409.

[125] Hosokawa, H. Higashi, H. Nakamura, T. Kusumoto, *Appl. Phys. Lett.,* 67 (1995) 3853.

[126] S. Lamansky, P. Djurovich, D. Murphy, F. Abdel-Razzaq, H.-E. Lee, C. Adachi, P.E. Burrows, S.R. Forrest, M.E. Thompson, *J. Am. Chem. Soc.,* 123 (2001) 4304.

[127] F.-I. Wu, H.-J. Su, C.-F. Shu, L. Luo, W.-G. Diau, C.-H. Cheng, J.-P. Duan, G.-H. Lee, *J. Mater. Chem.,* 15 (2005) 1035.

[128] Tokailin, M. Matsuura, H. Higashi, C. Hosokawa, T. Kusumoto, *Proc. SPIE,* 1910 (1993) 38.

[129] C. Adachi, T. Tsutsui, saito, *Appl. Phys. Lett.,* 56 (1990) 799.

[130] Y. Hamhda, T. Sano, M. Fujita, T. Fujii, Y. Nishio, K. Shibata, *Jpn. J. Appl. Phys., Part,* 2 35 (1993) L511.

[131] Y. Hamada, T. Sano, M. H. Fujii, Y. Nishio, H. Takahaski, K. Shibata, *Jpn. J. Appl. Phys.,* Part 2 35 (1996) L1339.
[132] Y. Hamada, C. Adachi, T. Tsutsui, S.Saito, *Jpn. J. Appl. Phys.,* Part 1 31 (1992) 1812.
[133] M. Strukelj, R. Jordan, and A. Dodabalapur, *J. Am. Chem. Soc.,* 118 (1996) 1213.
[134] Kido, W. Ikeda, M. Kimura, K. Nagai, *Jpn. J. Appl. Phys.,* Part 2 35 (1996) L349.
[135] C. W. Tang, S. A. Van Slyke, C. H. Chen, *J. Appl. Phys.,* 65 (1989) 3610.
[136] C. Adachi, T. Tsutsui, S. Saito, *Appl. Phys. Lett.,* 55 (1989) 1489.
[137] Mort, G. Pfister, Photoelectronic Properties of Photoconducting Polymers, (Wiley, New York, 1982), Chap. 6 (1982) 215.
[138] B. M. Bahirwar, R. G. Atram, R. B. Pode, S. V. Moharil, Materials *Chem. and Phys.,* 106 (2007) 364.
[139] N. Thejo Kalyani, S.J. Dhoble; *Ren. Sus. Ene. Rev.,* 16 (2012) 2696– 2723.
[140] Zhao, S. Xie, S. Han, Z. Yang, L. Ye, T. Yang, *Synth. Met.,* 114 (2000) 251.
[141] Kido, K. Nagai, and Y. Okamoto, *Chem.Lett.,* 13: (1990)657.
[142] L. Segura, *Acta. Polym.,* 49(1998) 319.
[143] Gauiter, A. Lovin, J.M. Nunzi, A. Schalchli, J. J. Benattar, D. Vital, *Appl. Phys .Lett.,* 69(1996) 1071.
[144] H. Aziz, Z.D.Popovic,N.X.Hu, A. M. Hor, G.Xu, *Science,* 283(1999) 1900.
[145] M. Do, E. Han, Y. Niidome, M. Fujihira, *J. Appl. Phys.,* 76(1994)5118.
[146] J. Mc Elvain, H. Antoniadis, M. R. Hueschen, J. N. Miller, D. M. Roitman, J. R. Sheats, R.L. Moon, *J. Appl. Phys.,* 80 (1996)6003.

In: Organic Light-Emitting Diodes (OLED)
Editor: Douglas Rivera
ISBN: 978-1-63484-001-9

Chapter 3

SUPPRESSION OF THE EFFICIENCY ROFF-OFF: THE CHALLENGE TO OVERCOME FOR THE COMMERCIALIZATION OF OLEDS

Frédéric Dumur*
Aix-Marseille Université, CNRS,
Institut de Chimie Radicalaire ICR, Marseille, France

ABSTRACT

Since the early work of Tang and Van Slyke, Organic Light-Emitting Devices (OLEDs) attract broad attention of researchers due to their potential uses in flat panel displays and solid-state lighting. Some OLEDs proponents even believe that white OLEDs (WOLEDs) will soon replace the existing fluorescent and filament lamps for general lighting application. At present, OLEDs still suffer from an insufficient lifetime and the transfer of this technology to the mass market is another story. Doping the emissive layer with fluorescent or phosphorescent dyes is widely practiced for devices fabrication. To reach high efficiencies, a key requirement is that all electrically generated excitons must be harvested for light emission. However, even if emitters and especially metal complexes are usually doped into suitable host materials, quantum efficiency of phosphorescent OLEDs still tends to decrease at high current density because of competitive factors such as concentration quenching and triplet-triplet annihilation. The efficiency roll-off i.e., the reduction of the quantum efficiency with increasing the current density is nearly general and constitutes the most significant problem facing OLEDs. This chapter aims at reporting the last strategies that have been developed to address this issue. Especially, a systematic understanding of the structure–property–performance relationships is proposed.

---

* E-mail: frederic.dumur@univ-amu.fr

INTRODUCTION

In recent decades, Organic Light-Emitting Devices (OLEDs) have garnered great attention due to their potential uses as flat and/or solid-state light sources for next generation energy-saving and lightweight lighting or as backlight sources for full-panel display applications [1]. In order to reduce the energy consumption of OLEDs, highly emissive emitters are required and this necessity constituted an incentive for researchers to further develop new emitters. In this aim, heavy metal complexes were identified as the best candidates. Indeed, by the strong spin–orbit coupling existing within heavy-metal complexes, singlet and triplet states can be both harvested for light emission, enabling to achieve nearly 100% Internal Quantum Efficiency (IQE) [2]. A fourfold efficiency increase compared to fluorescent materials can thus be obtained with phosphorescent materials [3]. If the research on OLEDs originally focused on platinum, [4] osmium [5] and ruthenium [6] complexes as emitters, rapidly, iridium complexes were identified as more promising candidates and these complexes remains undoubtedly the most widely studied [7]. As main figure of merit, lifetime of the iridium excited states is sufficiently short to minimize the triplet–triplet (T–T) annihilation which is a key adverse factor for incorporating phosphorescent emitters in OLEDs [8]. To date, most of OLEDs are designed with neutral complexes. However, recently, M. E. Thompson and coworkers tried to develop OLEDs comprising soft salts as emitters [9]. In this approach, both cationic and anionic partners of the soft salt can contribute to the light emission [10]. Besides, the scope of this strategy is limited to solution-processed devices, the molecular weight that can reach soft salts impeding all sublimation. Whatever the emitters are and the approach that is developed (use of neutral complexes or soft salts), the efficiency roll-off i.e., the reduction of the quantum efficiency with increasing the current density, is nearly general and constitutes the most significant problem facing organic light-emitting devices [11]. This is the most essential factor for OLEDs to be usable in daily applications since aim is clearly to realize high efficiency at high brightness with a reduced energy consumption. Several factors have been identified to explain the decrease of the device efficiency with the current density: triplet-triplet (T-T) annihilation caused by the long lifetime of the triplet excitons which strongly increases the probability for bimolecular recombination but also by the space charge accumulation caused by a narrow exciton recombination region, [12] dissociation of excitons induced by the electric field, [13] exciton formation in the hole transporting layer, [14] triplet-polaron annihilation, [15] loss of charge balance with increasing the current density [16] or exciplex emission [17]. In most phosphorescent OLEDs (PhOLEDs), triplet-triplet annihilation is the main cause of reduced current efficiency at high brightness. If this phenomenon is difficult to suppress, it can however be strongly limited in the region of interest of manufacturers, namely from 100 to 10 000 cd/m². Understanding of exciton dynamics is thus crucial to overcome this problem [18]. Several approaches have been envisioned to address this issue, all based on the convenient choice of the host material. Indeed, due to their long living excited states, transition metal complexes must be dispersed in the emissive layer (EML) at a low doping level into a host matrix, the low dopant concentration being required to suppress the concentration quenching and triplet–triplet annihilation [19]. As the host matrix occupies the majority of the emissive layer, it thus strongly influences the overall device performance and the host material is as important as the phosphorescent dopant is. Therefore, a great deal of effort has been directed at replacing the usual unipolar hole-transporting hosts [20] by bipolar materials [21a,21b]. Recent results obtained in this field definitely proved bipolar materials to lead to much higher device efficiencies and slower roll-off by favoring a more balanced charge transportation. Development of new triplet emitters with higher electron-transport ability can also be developed [21c]. With aim at reducing the efficiency roll-off, blue PhOLEDs constitute a model for the research in this field [22]. Indeed, compared to the very efficient green and red devices, blue PHOLEDs are especially challenging to fabricate by the difficulty of designing stable and highly emissive wide bandgap emitters as well as in finding suitable host materials with shallow triplet energies. If derivatives of carbazole [23] and silicon containing compounds [24] exhibit high triplet energy levels suitable to efficiently confine the triplet excitons on the dopant molecules, these wide bandgap materials often suffer from thermal and morphological instability. A tradeoff has thus to be found and it is difficult to meet the requirements of shallow LUMO, high triplet energy level and thermal stability. A proper device architecture design can also readily improve device efficiency. Several attempts were notably made to regulate the exciton concentration and/or broaden the exciton recombination zone by developing devices with double emissive layers, emissive layers separated by one or several interlayers or by simply mixing two co-hosts at different ratios to adjust the hole and electron transportation of the EML. Other groups worked on the simplification of the device architecture to facilitate the manufacturing process, even if this research remains at present marginal [25]. This chapter aims at reporting the different strategies that are currently developed to suppress the efficiency roll-off or at least to limit its negative impact on the electroluminescence performances.

1. Reduction of the Space Charge Accumulation by Use of Bipolar Hosts

An efficient strategy to reduce the efficiency roll-off has recently been developed for blue PhOLEDs [26]. As mentioned in the introduction, blue PhOLEDs constitute an excellent model for addressing the roll-off issue. Indeed, these OLEDs suffer from the strongest decrease of the efficiency with increasing the current density among all investigated OLEDs as a result of their intrinsic wide bandgap. In this study, the different materials were carefully selected to reduce the energy barrier between layers and thus to make the injection of holes and electrons inside the emitting layer easier. Indeed, the operating voltage of an OLED can be defined as the sum of the voltage corresponding to the photon energy of the emitted light and an over-potential corresponding to the energy required to inject electrons and holes from the two electrodes to the emitting layer. Ideally, the driving voltage could be reduced to the photon energy of the emitted light if the charge-injection barrier from the charge-transport process to the EML becomes negligible, so that no energy barrier exits (see Figure 1).

In this work, authors tried to develop these ideal OLEDs with an energy barrier-free structure. While using 9-(3-(9*H*-carbazole-9-yl)phenyl)-3-(dibromophenylphosphoryl)-9*H*-carbazole (mCPPO1) as the host material and *N,N'*-dicarbazolyl-3,5-benzene (mCP) as the hole injection material, the energy barrier for hole injection could notably be reduced to 0.03 eV. Similarly, energy barrier for electron injection between diphenylphosphine oxide-4-(triphenylsilyl)phenyl (TSPO1) acting as the exciton blocking layer and mCPPO1 used as host could be minimized to 0.12 eV. To end, the host and the dopant (*bis*((3,5-difluoro-4-cyanophenyl)pyridine)iridium (III) picolinate (FCNIrpic)) (see Figures 2 and 3) were selected in such a way that the energy differences between highest occupied molecular orbitals (HOMOs) and lowest unoccupied molecular orbitals (LUMOs) were similar, favoring a balanced hole and electron injection within the emissive layer. Using this strategy, charge accumulation at both interfaces could be drastically reduced.

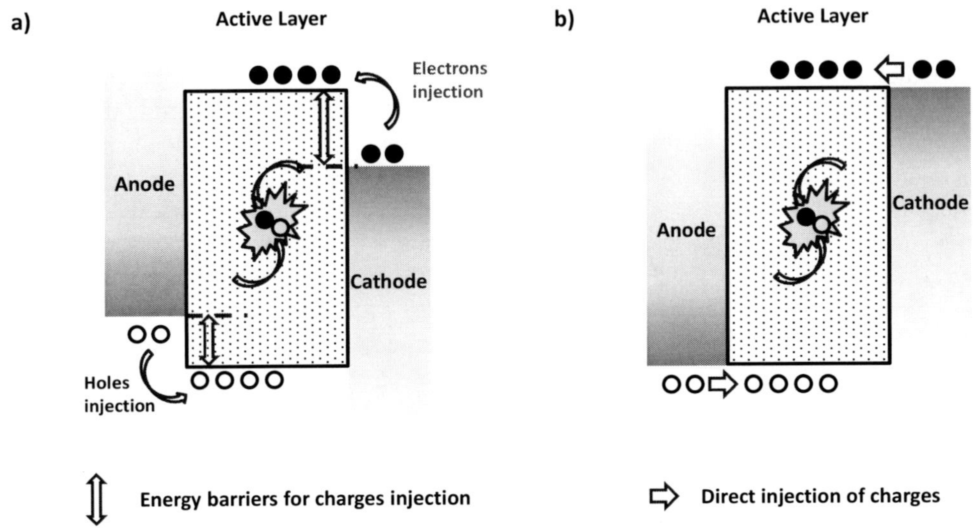

Figure 1. Energy diagram of an OLED classically studied in laboratories (a) and the ideal structure of a barrier-free device (b).

When using the optimum dopant concentration of 3 wt%, external quantum efficiencies (EQEs) of 21.7% and 20.6% were respectively obtained at 500 and 1000 cd/m². These values constitute the best results ever obtained for deep blue PhOLEDs with y chromaticity coordinates below 0.20. In particular, quantum efficiency could be kept higher than 20% until 1000 cd/m², a result which was unprecedented in 2008 [2c,27]. The design of highly emissive blue PhOLEDs requires the development of wide bandgap hosts. It can only be achieved if the electronic delocalization within the host molecule is limited i.e., the π-conjugation is significantly reduced. As drawback, the charge transport properties of these materials are adversely affected by this reduction of conjugation [28]. However, a few groups succeeded to design bipolar hosts combining charge transport ability and high triplet energy level. As first example, two bipolar hosts (2,6-*bis*(3-(carbazol-9-yl)phenyl)pyridine (26DCzPPy) and 3,5-*bis*(3-(carbazol-9-yl)phenyl)pyridine (35DCzPPy)) comprising the electron donating carbazole moiety well-known for its high triplet energy (E_T), [29] and the electron withdrawing pyridine also well-reported to exhibit a high electron affinity were combined within a unique structure [30]. The bipolar character of the new hosts was demonstrated while comparing the emission spectra of 4,4'-*bis*(9-carbazolyl)biphenyl (CBP)-based devices and 26DCzPPy- and 35DCzPPy-based devices using iridium (III) *bis*(4,6-(difluorophenyl) pyridinato-*N*,$C^{2'}$)picolinate (FIrpic) as the blue-green phosphorescent dopant (11 wt%) and by introducing luminescent markers on both sides of the EML, namely *N*,*N'*-*bis*((1-naphthyl)-*N*,*N'*-diphenyl-1,1'-biphenyl-4,4'-diamine (NPB) and *tris*(8-hydroquinolinato)aluminum

(Alq_3) that both emits at different wavelengths. Interestingly, the emission of Alq_3 dominated the emission spectra of CBP-based devices, indicating the charge recombination to occur at the CBP/Alq_3 interface. In contrast, emission of the two markers was detected for devices built up with 26DCzPPy and 35DCzPPy, indicating the charge recombination to occur in the whole EML, and not only at one interface.

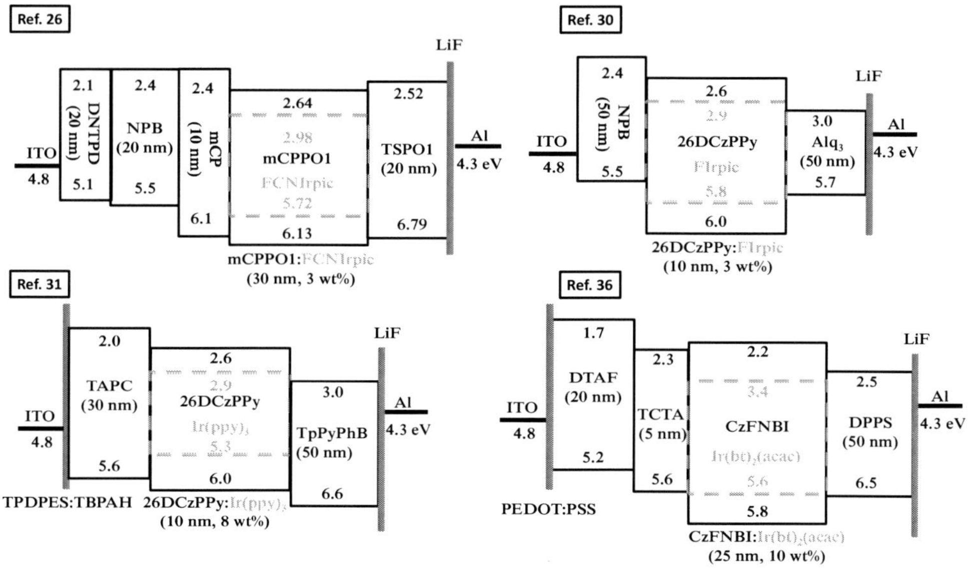

Figure 2. Representative energy diagrams of OLEDs designed to limit space charge accumulation and reported in references [26], [30], [31] and [36].

Based on the emission intensity, 26DCzPPy was determined as the host exhibiting the more balanced hole and electron injection and transport of the two molecules. An EQE of 24.3% was notably reached at the practical brightness of 100 cd/m^2, corresponding to a power efficiency (PE) of 46.1 lm/W. Severity of the efficiency roll-off behavior was significantly alleviated compared to previous blue PhOLEDs [27].

Thereby, EQE and PE only rolled off to 22% and 34 lm/W respectively at the brighter emission of 1000 cd/m^2. In the continuation of this work, a more systematic study on the effect of the nitrogen atom orientations in host materials comprising heterocyclic rings was carried out [31]. Choice of the central heterocyclic arylenes, the number of nitrogen atoms and their orientations was determined as the main factor influencing the HOMO and the LUMO levels of the host materials. Once again, 26DCzPPy was determined as the host exhibiting the more balanced hole and electron transportation. Using Ir(ppy)$_3$ as the dopant, the remarkable PE of 108 lm/W was obtained, corresponding to an EQE of 26.5% at 10 cd/m^2. Comparison between a series of seven hosts only comprising one or two nitrogen atoms in the central ring clearly evidenced a suppression of the efficiency roll-off for all hosts based on pyrimidine or pyrazine cores, even if lower performances than that obtained with 26DCzPPy were however determined.

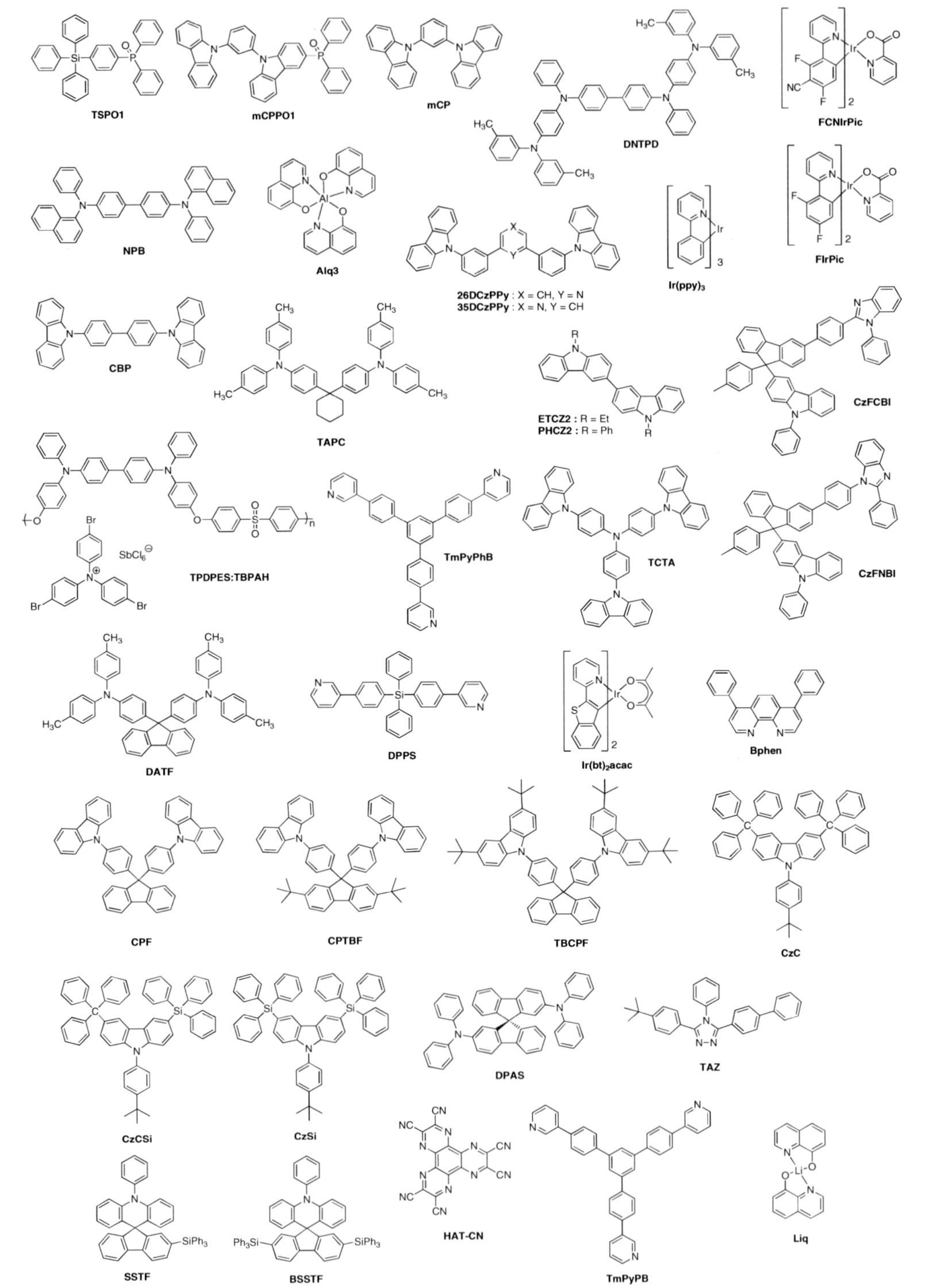

Figure 3. (Continued).

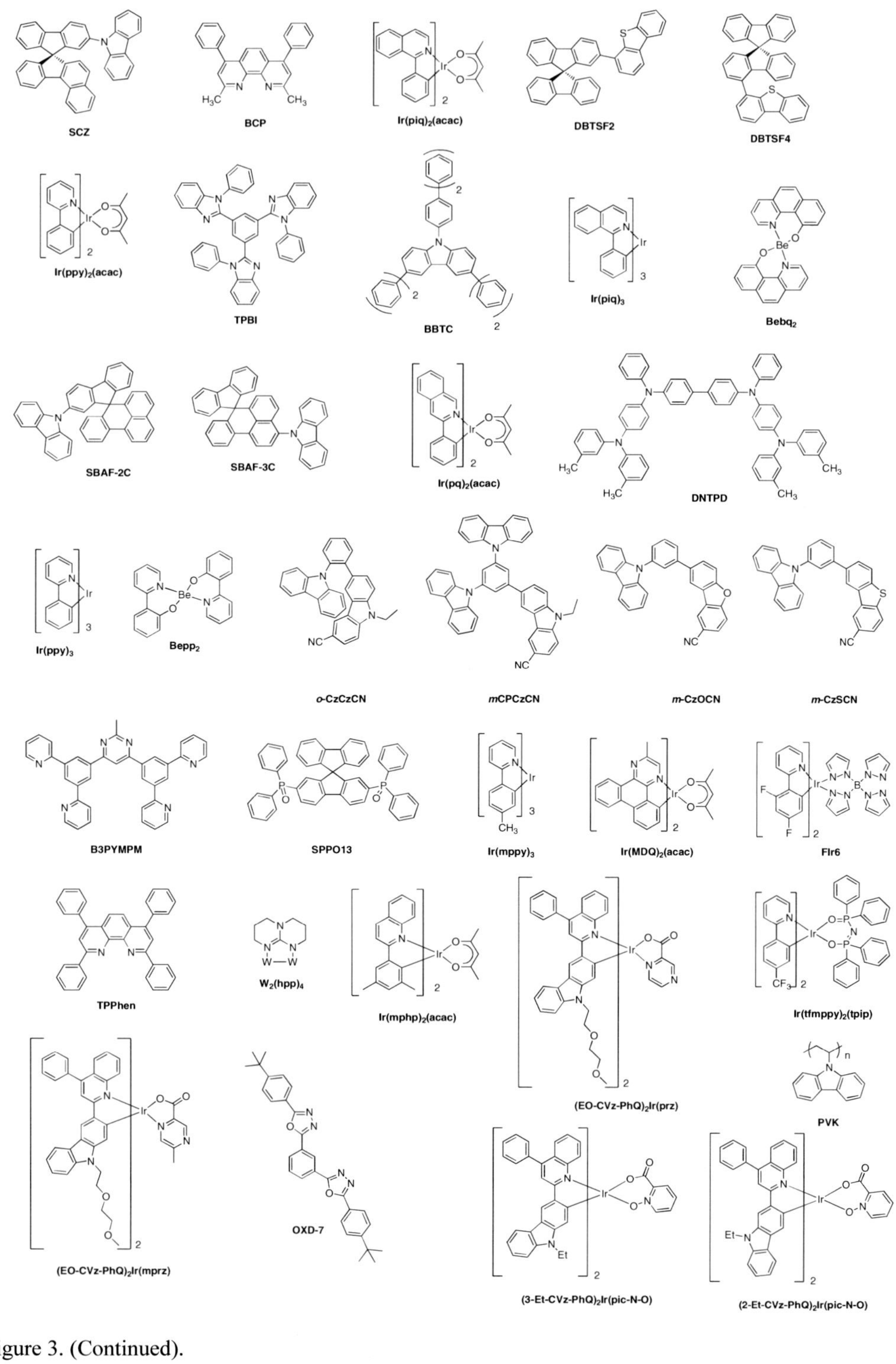

Figure 3. (Continued).

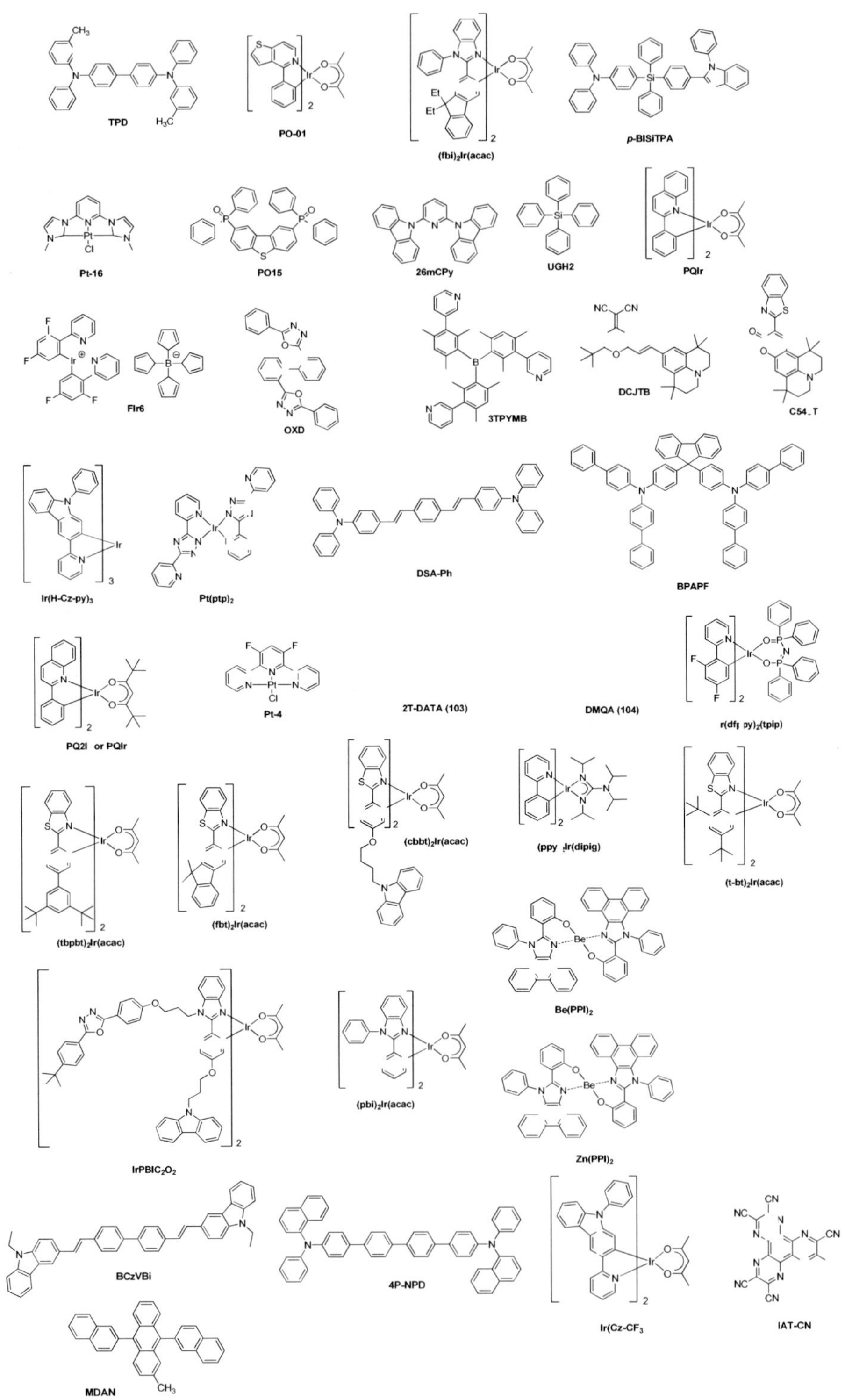

Figure 3. Chemical structures of materials used to fabricate OLEDs and cited in this chapter.

This trend was confirmed with benzimidazole-based materials, [32] and three-carbazole-armed host materials [33]. Carbazole is a structure that has been the elemental building block for the design of a wide range of materials. Several remarkable structures have been designed, addressing the roll-off issue as hole-transport materials or as bipolar hosts. To illustrate the first point, 9,9'-diethyl-9*H*,9'*H*-3,3'-bicarbazole (ETCZ2) and 9,9'-diphenyl-9*H*,9'*H*-3,3'-bicarbazole (PHCZ2) can be cited as successful structures to suppress the efficiency roll-off [34]. Using these molecules as hole-transport layers (HTL), the roll-off was limited to only 8% for PHCZ2-based devices in the 1000-10000 cd/m^2 range and 14% for ETCZ2-devices, contrarily to 21% observed for the reference devices using NPB. Similarly, DCDPA-based devices (where DCDPA stands for 3,5-di(9*H*-carbazol-9-yl)-*N*,*N*-diphenylaniline) furnished red, green and deep-blue PhOLEDs with efficiency roll-off lower than 10%, contrarily to mCP-based devices that exhibited a roll-off close to 30% in the same luminance range [35]. To illustrate the second point, investigation of the structural topology and connectivity between fluorene-bridged benzimidazole and carbazole moities enabled to subtly modulate the physical properties of tailor-made bipolar hosts [36]. Two materials were studied, namely 9-phenyl-3-(3-(4-(1-phenyl-1*H*-benzo[*d*]imidazol-2-yl)phenyl)-9-(*p*-tolyl)-9*H*-fluoren-9-yl)-9*H*-carbazole (CzFCBI) and 9-phenyl-3-(3-(4-(1-phenyl-1*H*-benzo [*d*]imidazol-2-yl)phenyl)-9-(*p*-tolyl)-9*H*-fluoren-9-yl)-9*H*-carbazole (CzFNBI) which vary by the N- or the C-connectivity between the two partners. Hosts perfectly matching the energy level of the emitting layer and with those of the neighboring layers were thus prepared (see Figure 2). This excellent alignment enabled to severely reduce the roll-off of OLEDs fabricated in a really common device configuration. Especially, the N-connectivity imparted a high triplet energy level to CzFNBI ensuring its utilization as a universal host for blue-to-red phosphors. While examining the performances of blue-to-red PhOLEDs, highest EQE (19.6%) was obtained with *bis*(2-phenylbenzothiazolato)(acetylacetonato)iridium (III) ($Ir(bt)_2(acac)$) as the yellow dopant. Lower triplet energy of CzFCBI limited its host ability to green-to-red dopants. If the scope of this materials is more limited, high-performance devices can however be fabricated with this material as host, as illustrated with the CE and PE of 74.4 cd/A, 75 lm/W and the remarkable maximum EQE of 20% achieved with $Ir(ppy)_2(acac)$ as the green dopant. Still in the same spirit, fluorene/carbazole hybrids were developed as bipolar hosts for blue organic electrophosphorescent devices [37]. With regard to hinder self-quenching of the triplet excitons in the EML issued from strong bimolecular interactions between phosphorescent dyes, a method commonly used consists in introducing bulky, rigid, and inert peripheral substituents on the dopants [38]. Here, the opposite strategy was developed and two sterically hindered host materials 9,9-*bis* [4-(3,6-di-*tert*-butylcarbazol-9-yl)phenyl]fluorine (TBCPF) and 9,9-*bis* [4-(carbazol-9-yl)phenyl]-2,7-di-*tert*-butylfluorene (CPTBF) were prepared. High glass transition temperatures (T_g >160°C) and high E_T (>2.8 eV) were obtained for the two molecules. When tested as hosts for blue PhOLEDs, the suppression of concentration quenching by use of these sterically hindered hosts was clearly evident. For comparison, reference devices were prepared with the unsubstituted 9,9-*bis*[4-(9-carbazolyl)phenyl]fluorene (CPF) host. As first evidence, extremely high dopant concentration could be used with the bulky host materials without reduction of current efficiency (CE). Thus, for TBCPF-based devices doped with FIrpic at 15 wt%, the peak CE of 10.9 cd/A obtained at 15 wt% could be even increased to 12.7 cd/A at 20 wt% dopant concentration, evidencing the suppression of the concentration quenching in this host. This unusual behavior was not confirmed with CPTBF that differs from TBCPF by the number and the position of the bulky substituents. As appealing feature, devices showed little efficiency roll-off. Less than 7% drop in the CE at 1000 A/m² was found for TBCPF at 15 wt% dopant concentration, while more than 26% roll-off was observed with CPF-based devices. Silicon and carbazole-containing compounds were also visited as potential hosts for blue PhOLEDs and a series of triphenylsilyl- and trityl-substituted carbazole-based materials was proposed [39]. To destabilize the LUMO levels and to reduce the π-conjugation, triphenylsilyl (-$SiPh_3$) and trityl (-CPh_3) side groups were connected to the carbazole central core adopting a nonconjugated substitution strategy. Contrarily to other silicon-based materials that suffer from an insufficient thermal and morphological stability, the three bipolar materials developed in this study, i.e., 9-(4-*tert*-butylphenyl)-3,6-*bis*(triphenylsilyl)-9*H*-carbazole (CzSi), 9-(4-*tert*-butylphenyl)-3,6-ditrityl-9*H*-carbazole (CzC), and 9-(4-*tert*-butylphenyl)-3-(triphenylsilyl)-6-trityl-9*H*-carbazole (CzCSi) exhibited glass transition temperatures ranging from 131 to 163°C, and the decomposition temperatures varied from 388 to 393°C. As specificity, charge transport ability of these materials showed a rather significant dependence to the substitution pattern. Hole mobility of CzC (~5×10^{-4} cm^2/Vs) was thus found to be one order of magnitude higher than that of CzSi (~5×10^{-5} cm^2/Vs) whereas the hole mobility of CzCSi was in between. While examining the EL performances, devices fabricated with CzSi exhibited low efficiency roll-off compared to that observed for devices using CzC or CzSi as hosts, with an EQE above 12% at 100 cd/m^2 and 9% at 1000 cd/m^2. On the opposite, CzC and CzCSi-based devices suffered from severe efficiency roll-off. EQE of CzC-based devices notably dropped from 7% at 100 cd/m^2 to 0.6% at 1000 cd/m^2, corresponding to a drop of the EQE higher than 92%. Higher performances of CzSi devices over the other devices was notably ascribed to the lower hole mobility of the host compared to the two others, favoring the formation of a broad exciton distribution zone over the whole emitting layer. Contrarily to the two other hosts that exhibit a high hole mobility and thus favor the excitons accumulation at the EML/ETL interface, the distribution of excitons over the whole EML thus severely limit the excitons quenching by triplet-triplet (T-T) annihilation for CzSi devices.

Silicon-based materials comprising spiro-annulated fluorenes and triphenylamines units also fullfiled the different requirements of impeding the reverse energy transfer from the guest to the host, forming stable amorphous films and exhibiting HOMO and LUMO levels perfectly fitting with those of the adjacent layers [40]. Remarkably, devices fabricated with the dissymmetric 10-phenyl-2'-(triphenylsilyl)-10*H*-spiro [acridine-9,9'-fluorene] (SSTF) exhibited slightly higher performances (44.0 cd/A, 41.5 lm/W) than that obtained with the symmetric 10-phenyl-2',7'-*bis*(triphenylsilyl)-10*H*-spiro [acridine-9,9'-fluorene] (BSSTF) (40.9 cd/A, 35.2 lm/W). These results are significantly higher than those reported in previous works [23b,41]. Efficiency roll-off was exceptionally low for SSTF-based devices and limited to 4.8% at the practical brightness of 1000 cd/m². Parallel to carbazoles, research on spirofluorene-containing hosts has also been extremely dynamic during the last decade. The next spirofluorene-benzofluorene-based molecule i.e., 9-(spiro[benzo[*c*]-fluorene-7,9'-fluorene]-2'-yl)-9*H*-carbazole (SCZ) that was developed for the red dopant $Ir(piq)_2(acac)$ is one illustration of the chemical engineering around spirofluorenes [42]. High efficiency of OLEDs is closely related to the host-dopant energy level alignment for efficiently confining excitons on the dopant. Especially, the dopant should have its HOMO and LUMO levels within the bandgap of the host material and this condition was clearly verified with $Ir(piq)_2(acac)$. Surprisingly, efficient devices could only be obtained at extremely low dopant concentration. Decrease of the device performance with increasing the dopant concentration was assigned to both holes and electrons trap by the dopant, high energy barriers between the HOMOs and LUMOs of the host and the guest being observed. After optimization, maximum quantum efficiency was obtained at 1% dopant concentration. However, due to the low concentration, incomplete energy transfer was observed from the host to the guest, resulting in both the dopant and the host emission. Conversely, even if lower performances were obtained by increasing the dopant concentration, no efficiency roll-off was observed, whatever the dopant concentration is, evidencing the balanced charge transport of the host. This last remark must be tempered, however, by the low EQE discussed in this article (maximum EQEs from 2 to 8%, depending of the dopant concentration). Combination of spirofluorene was also extended to dibenzothiophene to furnish the new bipolar hosts (4-(9,9'-spirobi[fluoren]-2-yl)dibenzo[*b,d*]thiophene (DBTSF2) and 4-(9,9'-spirobi[fluoren]-4-yl)dibenzo[*b,d*]thiophene (DBTSF4)) tested in green PHOLEDs [43]. In this series, the linkage adopted between the two moieties had a significant impact on the triplet energy level of the resulting hosts and an energy difference of 0.33 eV could be determined between the T_1 states of the two regioisomers. While testing a green dopant, best devices achieved a maximum CE of 62.0 cd/A, a maximum PE of 45.9 lm/W and a maximum EQE of 16.3% with the host exhibiting the highest T_1 energy (DBTSF4). Achieving the best alignment of the energy levels of the host with those of the proximal layers was the main focus of the next study [44]. As already mentioned, high energy barrier between HTL/EML and/or EML/ETL results in charges trapping by the dopant, what is detrimental for device performances. Considering the HTL and ETL materials that are currently at disposal for designing OLEDs, synthesis of bipolar hosts with HOMO and LUMO levels that will more efficiently fit with those of the adjacent layers was developed. Synthesis of such a molecule was achieved with the synthesis of 3,6-*bis*-biphenyl-4-yl-9-[1,1',4',1'']terphenyl-4-yl-9*H*-carbazole (BBTC), fulfilling all the above requirements. T_g of BBTC was measured at 148°C, which is relatively high compared to that reported for the other hosts also comprising carbazole units in their backbone [23b, 45]. These results were only recently overcome with carbazole-based hosts exhibiting T_gs as high as 174 and 168°C [46]. To investigate the host ability of BBTC, the following device structure was investigated to fabricate red PhOLEDs with two different dopants: ITO/ MoO_3 (10nm)/ NPB (50nm)/ BBTC doped with $Ir(piq)_3$ or $(piq)_2Ir(acac)$ (8 wt%, 30 nm)/ BCP (10 nm)/ $Bebq_2$ (40 nm)/ LiF (0.5 nm)/ Al. Best performances were obtained with $Ir(piq)_2(acac)$. Notably, the exceptional brightness of 82 625 cd/m² was achieved at 14.3 V. A maximum EQE of 19.3% and a maximum CE of 16.4 cd/A were also obtained at 100 cd/m². A current efficiency roll-off of only 20% was observed between 100 and 10 000 cd/m², which is relatively low, considering the brightness range. These results outperform in efficiency the best results ever reported in the literature for red PhOLEDs [47]. High efficiency and reduced roll-off were assigned to the relatively high HOMO energy level of BBTC compared to that of conventional host materials, facilitating the hole injection in the EML. Suppression of the triplet exciton quenching processes through the enhanced electron and hole balance of BBTC was also suggested to explain the high device performance.

Effectiveness of carbazole-based host materials for blue PHOLEDs due to the high E_T of the carbazole unit has already been evoked earlier in this chapter. Jointly, spirobifluorene derivatives are reported to exhibit high T_gs and excellent morphological stabilities [48]. By combining spirobenzofluorene with carbazole units within a unique hybrid material, electron and hole transport properties were respectively attended from each moiety. Two carbazole-containing spirobenzoanthracene-type materials, namely 9-(spiro- [benzoanthracene-7,9'-fluorene]-2'-yl)-9*H*-carbazole (SBAF-2C) and 9-(spiro [benzoanthracene-7,9'-fluorene]-3-yl)-9*H*-carbazole (SBAF-3C) were thus prepared and tested as new hosts for the design of red PhOLEDs [49]. In these two structures, the hole-transporting carbazole moiety was introduced at the 2- or 3-positions of the central electron-transporting spirobenzofluorene unit. These unique combinations were supposed to furnish bipolar materials with balanced electrons and holes transportation. However, the mismatch between the energy levels of the two host materials with those of the red dopant $Ir(pq)_2acac$ resulted in an emission peak from the host irrespective to the dopant concentration furnishing orange PhOLEDs due to the host contribution. Among the two hosts, SBAF-3C exhibited the lowest performances, assigned to its lower triplet state and resulting in a less efficient energy transfer on the dopant. With the configuration of ITO/ DNTPD (60 nm)/ NPB (30 nm)/ SBAF-2C:$Ir(pq)_2acac$ (30 nm, 3 wt%)/ BCP (5 nm)/ Alq_3 (20 nm)/ LiF (1 nm)/ Al (with DNTPD : *N,N'*-di(1-naphthyl)-*N,N'*-diphenylbenzidine and BCP : 2,9-dimethyl-4,7-diphenyl-1,10-phenanthroline), a maximum EQE of 8.93%, a peak CE and PE of 14.7 cd/A and 7.7 lm/W were only obtained as a result of the inefficient energy transfer from the host to the dopant. In this study, we can clearly regret the use of a red dopant with inadapted energy levels to investigate these two new materials.

To date, *bis*[2-(2-hydroxyphenyl)pyridine]beryllium ($Bepp_2$) is the host with which PhOLEDs exhibiting the lowest efficiency roll-off have been reported [50]. It is worth noting that this host is a bit apart from the others since this is not an organic molecule but a metal complex. And this new approach seems to be promising for the future with respect to the results obtained by these authors. Only an efficiency drop of 8% could be observed over the brightness of 10 000 cd/m^2. To substantiate this argument, maximum CE and PE as high as 62.5 cd/A and 51.0 lm/W and an EQE of 19.8% were respectively obtained while doping the EML with 2 wt% $Ir(ppy)_3$. High efficiency of PhOLEDs clearly indicated the good confinement of excitons inside of the emissive layer and assigned to the shallow hole trapping existing in this host-guest system. Notably, the 0.4 eV hole barrier between $Bepp_2$ and $Ir(ppy)_3$ facilitated the transport of holes within the EML and limited hole trapping at phosphorescent guest molecules.

To end this part, the series of five carbazole-containing bipolar hosts (6-(3-(9H-carbazol-9-yl)phenyl)-9-ethyl-9*H*-carbazole-3-carbonitrile (*m*-CzCzCN), 6-(2-(9*H*-carbazol-9-yl) phenyl)-9-ethyl-9*H*-carbazole-3-carbonitrile (*o*-CzCzCN), 6-(3,5-di(9*H*-carbazol-9-yl) phenyl)-9-ethyl-9*H*-carbazole-3-carbonitrile (*m*CPCzCN), 8-(3-(9*H*-carbazol-9-yl) phenyl) dibenzo[*b,d*]furan-2-carbonitrile (*m*-CzOCN), 8-(3-(9*H*-carbazol-9-yl) phenyl)dibenzo [*b,d*]thiophene-2-carbonitrile (*m*-CzSCN)) that is now described has been specifically designed for blue PhOLEDs. This series of materials differ from the former ones by the substitution pattern where electron withdrawing CN groups have been introduced on the carbazole, dibenzofuran or dibenzothiophene unit [51]. It is worth noting that bipolar hosts with cyano substituents have not been so extensively studied in OLEDs [52]. While evaluating the five bipolar hosts in OLEDs with the simple structure of ITO/ PEDOT:PSS (40 nm)/ 1,1-*bis*[(di-4-tolylamino)phenyl]cyclohexane (TAPC) (20 nm)/ host:8wt% FIrpic (30 nm)/ TmPyPB (40 nm)/ LiF (1 nm)/ Al, *o*-CzCzCN-based devices exhibited the best performances, with a maximum brightness of 24 300 cd/m^2 at 15 V, a peak CE of 43.9 cd/A, a maximum PE of 23 lm/W and a maximum EQE of 21%. Especially, peak CE obtained with *o*-CzCzCN are close to the best data reported so far (46 cd/A, [53] 49.4 cd/A [54] and

58.7 cd/A [55]) for Firpic-based PhOLEDs. Only a moderate roll-off was also detected for *o*-CzCzCN-based devices since, at a brightness of 1000 cd/m², the EQE only decreasing of 14% relative to the maximum value. From the numerous examples reported in this part, it can be clearly concluded that expending the recombination zone to the whole emissive layer is an effective strategy to address the roll-off issue. This goal was achieved by the convenient design of host materials. But other approaches were also investigated. Easier to handle than the design of new hosts by chemical engineering, the mixture of benchmark host materials proved to be another effective strategy.

2. Solving the Charge Carrier Imbalance in Emissive Layers by Use of Co-Hosts

As already mentioned, a proper device architecture design can readily increase the device efficiency. To improve the charge balance within the emissive layer, use of co-hosts is a new strategy that has recently been developed to reduce the triplet and polaron density and thus limit the efficiency roll off. This strategy is easier to implement since the two hole- and electron-transporting partners are not covalently linked, and that the bipolar character of the co-host can be rapidly adjusted during device elaboration by modifying their respective ratio. A rapid screening can be thus performed to achieve the right ratio. To date, only few examples of OLEDs making use of co-host systems were reported, only with a limited success in terms of reduction of the efficiency roll-off [56]. However, by the convenient choice of the hole and electron transporting materials, the transport property of the blend can be adjusted such as reaching the perfect balance between the transport of electrons and holes, greatly improving the EL efficiency. In this recent example using TCTA and *bis*-4,6-(3,5-di-3-pyridylphenyl)-2-methylpyrimidine (B3PYMPM) as co-hosts and $Ir(ppy)_2(acac)$ as the dopant, an extremely low roll-off could be evidenced [57]. Notably, if an EQE of 27.8% at 10 000 cd/m² could be obtained, EQE remained almost unchanged at 20 000 cd/m² and still reached the exceptional value of 26.0%, representing over than 95% of the maximum EQE. OLEDs also achieved the remarkable maximum PE of 124 lm/W, definitely evidencing the benefits of using co-hosts (See Table 1). Due to the high EQE and the almost quantitative conversion of electrons to excitons, this exceptional low roll-off was notably assigned to the suppression of the triplet-polaron quenching usually arising within the emissive layer, all polarons being used to generate excited states of the co-hosts or the dopant.

High power efficiency blue PHOLEDs were also obtained using another mixed host emitting layer made of TAPC and 2,7-*bis*(diphenylphosphoryl)-9,9'-spirobi[fluorene] (SPPO13) as the co-host materials and FIrpic as the blue phosphorescent dopant [58]. Notably, SPPO13 was selected for its electron transport properties and TAPC for its hole transport properties. Best ratio of TAPC:SPPO13 was determined as being 25:75, based on an extensive work of optimization. While using this co-host ratio, an EQE of 17.9% and a PE of 28.4 lm/W at 1 000 cd/m² were respectively obtained. High PE even at high luminance and thus the low roll-off was here again obtained as a result of the balanced charge transport and injection in the EML. Surrounding of the emissive layer and its interactions with HTL and/or ETL was examined in details for green PhOLEDs comprising a mixture of TAPC and TmPyPb as co-host [59]. In this remarkable study, a thorough optimization of the device structure (thickness, number of layers) was carried out. Best efficiency results were obtained for the bilayer structure ITO/ TAPC (10 nm)/ TAPC:TpPyPB:Ir(ppy)$_3$ (1:1, 5 wt%, 70 nm)/ LiF (0.5 nm)/ Al where a thin layer of TAPC has been introduced between the ITO electrode and the blend of EML. Peak CE and PE as high as 32.8 cd/A and 29.1 lm/W, an EQE of 13.8% were respectively obtained at 1000 cd/m^2. By introducing the TAPC interlayer between the electrode and the EML, exciton quenching process at the electrode/EML could be efficiently suppressed, enhancing the devices performance. To demonstrate it, device performances without the TAPC interlayer were reduced to 8.1 cd/A, 4.5 lm/W and 3.1%, for the CE, PE and EQE respectively. TCTA is a well-known material widely used as single host for phosphorescent dopants in OLEDs. To fabricate devices with a more balanced carrier injection, TCTA was also blended with different electron-injection aiding co-hosts. As first example, mixture of the hole-transporting TCTA with the electron-transporting SPPO13 provided a universal solution-processable bipolar co-host for blue, green and red inverted OLEDs [60]. Charge transport and recombination ability of the co-host could be finely tuned by changing the ratio of the two materials, providing an adequate tool to optimize the charge balance. An optimum ratio of 2:1 for SPPO13:TCTA was determined for all OLEDs. As appealing feature, EML films could be easily prepared by spin-coating, enabling to avoid the complicated step of a simultaneous co-evaporation of three materials. While examining device characteristics, maximum CE as high as 34.3, 56.9 and 12.0 cd/A could be achieved for blue, green and red devices, only rolling off to 30.8, 54.4 and 9.8 cd/A at 1000 cd/m^2. Efficiencies higher than that reported for blue OLEDs based on FIrpic and fabricated by thermal evaporation were notably obtained [61]. Previous to this study, the mixture of TCTA and TPBI used as co-hosts for Ir(ppy)$_3$ proved to be another effective mixture of materials for reducing the efficiency roll-off by expanding the recombination zone and reducing the probability of triplet-triplet exciton quenching (see Figure 4) [62]. If no precise values are provided in this article, EQE is however reported as remaining over 90% of its maximum at 20 000 cd/m^2. In the same field, the group of Chu and coworkers is developping since a decade various TCTA-based co-hosts, all efficiently addressing the roll-off issue. In one of their initial attempts, blend of TCTA with *m-bis*-(triphenylsilyl)benzene (UGH3) furnished a bipolar host for FIrpic and enabled to significantly reduce the efficiency roll-off at high luminance [63]. Remarkable peak EQE and PE of 21.6% and 44.9 lm/W were achieved for the blue PhOLEDs. While replacing FIrpic by *bis*(4',6'-difluorophenylpyridinato)tetrakis(1-pyrazolyl)borate (FIr6), a significant decrepancy of the device performance was observed. This unexpected behavior was understood as resulting from the ineffective ability of TCTA to efficiently host FIr6, consecutive to the mismatch of their respective energy levels [64]. The 0.4 eV lower-lying HOMO level of FIr6 relative to that of TCTA (5.7 eV) resulted in an inefficient emission of the dopant and a partial emission from TCTA. Following this initial study, the same authors studied another mixed host structure, still using FIrpic as the dopant [65]. By replacing UGH3 by 2-(diphenylphosphoryl)spiro-fluorene (SPPO1) as co-host for TCTA that exhibit similar energy levels to that of UGH3, the resulting combination furnished blue PhOLEDs with external quantum and power efficiencies values relatively close to that previously obtained with UGH3 (21.8% and 53 lm/W respectively (TcTa:SPPO1 = 3:7)). Reduced roll off in efficiency of TCTA:SPPO1-based devices compared to TCTA:UGH3-based devices was assigned to a broader recombination zone for TCTA:SPPO1-based devices resulting from a more balanced charge carrier injection/transport in this mixed host devices. If the roll-off could be significantly reduced in these last two strategies using UGH3 or SPPO1 as co-hosts for TCTA, the energy barrier between the emissive layer and the adjacent layers still remained, requiring the use of an additional driving voltage to overcome these energetic barriers, and thus favoring the charge accumulation.

Recently, another research group had the idea to simplify the device structures of Chu and coworkers by directly co-hosting the hole-transporting TCTA with the electron-transporting material used by Chu i.e., TmPyPb in order to fabricate high-efficiency blue PHOLEDs [66]. Importantly and to address the issue of extra electric field required for injecting holes and electrons in the EML, authors also used these two materials for the hole-transporting and electron-transporting layers, thus suppressing all energy barrier between the emissive layer and the two adjacent layers.

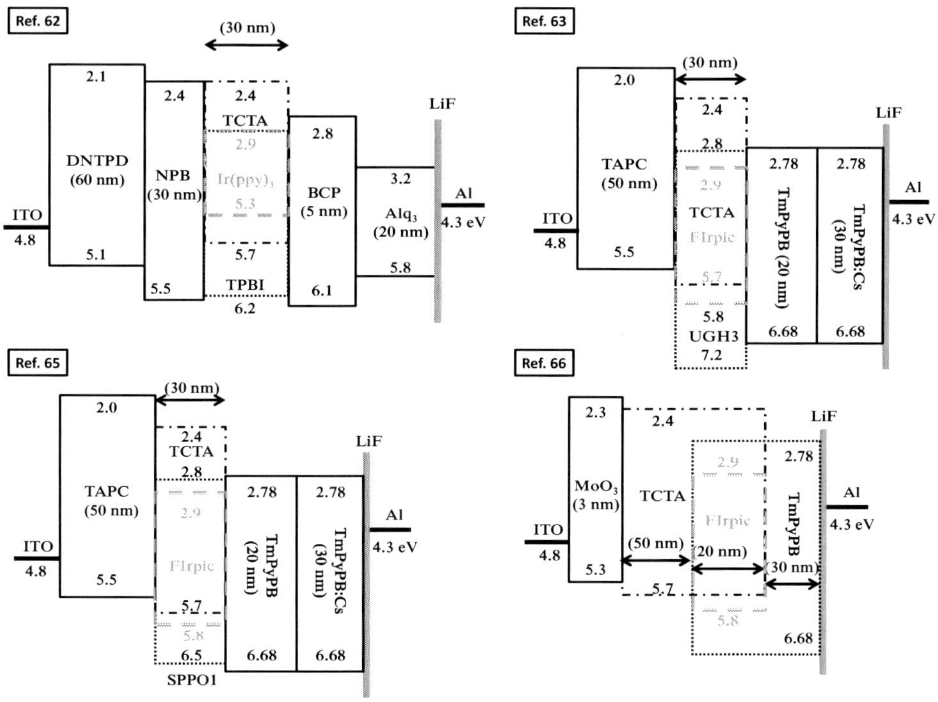

Figure 4. Energy diagrams of OLEDs designed with mixtures of co-hosts (taken from references [62], [63], [65] and [66]).

Especially, mixed-EML devices exhibited much higher efficiencies than that of the reference single-host EML counterparts. Improvement of the device performance was assigned to the stepwise energy-level architecture, leading to a significantly reduced energy-barrier to inject both hole and electron to inject within the EML. Notably, facile charge injection of the mixed-host devices was demonstrated by the relatively low turn-on voltage (2.6 V) as compared to the single-host devices (3.0 V). The maximum EQE and PE reached 20.4% and 55.4 lm/W respectively, far from the 18.6% and 47.6 lm/W of the TCTA-based devices and the 17.7% and 38 lm/W of the TmPyPb-based single host devices. While comparing the PE at the luminance of 100 cd/m² and 1000 cd/m², an efficiency roll-off value of only 18% was determined, less severe than that observed for TCTA-based devices (22%) and TmPyPb-based devices (35%). Enhancement of the device performance was assigned to a broader exciton recombination zone extending on the whole emissive layer and not only located at an EML side as commonly observed for single-host devices. High exciton density was thus avoided, reducing the triplet-triplet annihilation and exciton accumulation. A careful selection of the hole-transport material used to form the co-host is of crucial importance as it determines the ability of the resulting material to efficiently host the phosphorescent dopant. Notably, most of the hole-transport materials that are commonly used as hosts in EMLs are not suitable for phosphorescent dyes such as deep blue emitters as a result of their inappropriate triplet energy levels too low for an efficient energy transfer to the guest. This question was examined during the design of highly emissive green PhOLEDs in which mCP has been combined with TPBI to form the co-host of the EML [67]. mCP is a material commonly used as host for blue dopants which is characterized by its high E_T and its deep HOMO level (5.9 eV). However, use of this material as HTL is not so common since injection of holes from ITO in the deep HOMO is difficult. In this study, mCP (E_T = 2.9 eV) was selected over TCTA (E_T = 2.5 eV) for its higher triplet energy, enabling to host wide bandgap emitters. A new emitter, iridium (III) *bis*(2-(4-trifluoromethyl-phenyl)pyridine)-tetraphenylimidodiphosphinate Ir(tfmppy)$_2$(tpip) was selected as the green dopant for this study. Idea to mix TPBI with mCP aroused from the observation that the extremely low performances of mCP-based devices could originate from an excess of hole injection into the light emitting zone facilitated by the favorable alignment of the HOMO levels of both the HIL and the dopant. To address this issue and to achieve a better hole-electron balance within the EML, mCP and TPBI were blended. Electron transportation was also improved by adopting a novel electron transporting layer of higher electron transport ability, namely 2,4,7,9-tetraphenyl-1,10-phenanthroline (TPPhen) doped with tetrakis (1,3,4,6,7,8-hexahydro-2*H*-pyrimido [1,2-a]pyrimidinato)ditungsten (II) (W_2(hpp)$_4$) that replace the 4,7-diphenyl-1,10-phenanthroline (Bphen) layer of the mCP-based devices. A significant improvement of the device performances was obtained since the maximum CE increases of 29% compared to the same devices only comprising mCP as the single host and reached the value of 47.2 cd/A at 1000 cd/m^2. Only a 16% drop of the CE was observed at 10 000 cd/m^2 evidencing a more balanced charge transportation and a broader recombination zone.

A step further consisted in the mixture of two dopants in addition to the blend of co-hosts. If outstanding performances can be obtained with this approach, an industrial transfer of these results can't be reasonably envisioned with regards of the underlying technicity and material consumption it supposes. In the following study, all OLEDs were prepared by thermal evaporation. Notably, the co-host made of TCTA and B3PYMPM was co-doped with two phosphorescent dopants Ir(ppy)$_2$(acac) (iridium(III) *bis*(2-phenylquinoline) acetylacetonate) (red dopant) and Ir(mphq)$_2$(acac) (iridium(III) (*bis*(2-(3,5-dimethyl-phenyl)quinoline acetylacetonate) (green dopant)) to fabricate orange PhOLEDs [68]. The exciplex forming co-host was notably designed to avoid the large efficiency roll-off commonly observed for orange PhOLEDs and improves the color stability. By mixing the two materials, the resulting co-host showed a high triplet exciplex energy at ca. 2.5 eV. The device structure that was used is as follows: ITO/ TAPC (20 nm)/ TCTA (10 nm)/ TCTA:B3PYMPM:Ir(ppy)$_2$(acac):Ir(mphq)$_2$(acac) (30 nm, 8 wt%, 0.3 wt%)/ B3PYMPM (45 nm)/ LiF (0.7 nm)/ Al. The very high EQE of 25.4% was achieved, that only decreased to 21% at 10 000 cd/m^2. Examination of the transient electroluminescence of OLEDs evidenced a faster turn-on of the green emission, compared to the red one, clearly evidencing an energy transfer from the green to the red dopant (see Figure 5). To avoid a complete energy transfer from the green to the red dopant and maintaining a green emission, an extremely low red dopant concentration was used and optimum concentration was determined as being 0.3 wt%. Finally and parallel to this study, deposition of the same two dopants into two separated and adjacent EMLs using the same co-host mixture was investigated by the same authors and furnished in turn similar results [69].

The host blend approach, as part of an effective strategy for the design of PhOLEDs, was even extended to the mixture of three co-host materials, making the fabrication process relatively complicated if the emissive layer was not deposited by a solution based manufacturing process (see Figure 6) [70]. It also introduces additional complications directly related to material selection and device design. Hopefully, for this study, the solution process was selected for the deposition of the EML. Two new red dopants were tested i.e., *bis*[9-(2-(2-methoxyethoxy)ethyl)-3-(4-phenylquinolin-2-yl)-*9H*-carbazolato-N,C$^{2'}$]iridium 2-pyrazinecarboxylic acid (EO-CVz-PhQ)$_2$Ir(prz) and *bis*[9-(2-(2-methoxyethoxy)ethyl)-3-(4-phenylquinolin-2-yl)-*9H*-carbazo-lato-N,C$^{2'}$]iridium 5-methyl-2-pyrazinecarboxylic acid (EOCVz-PhQ)$_2$Ir(mprz). As host material, a blend of poly(vinylcarbazole) (PVK), *N,N'*-diphenyl-*N,N'*-(*bis*(3-methylphenyl)-[1,1-biphenyl]-4,4'-diamine (TPD) acting as the hole transport material and 1,3-*bis*[5-(4-*tert*-butylphenyl)-1,3,4-oxadiazole-2-yl]benzene (OXD-7) acting as the electron transport material was prepared in a 2:1:1 ratio. Especially, PVK was selected as hole-transporting polymer for its good solubility in common organic solvents, excellent film-forming properties, high glass-transition-temperature (200°C) and high E_T (3.0 eV) [71].

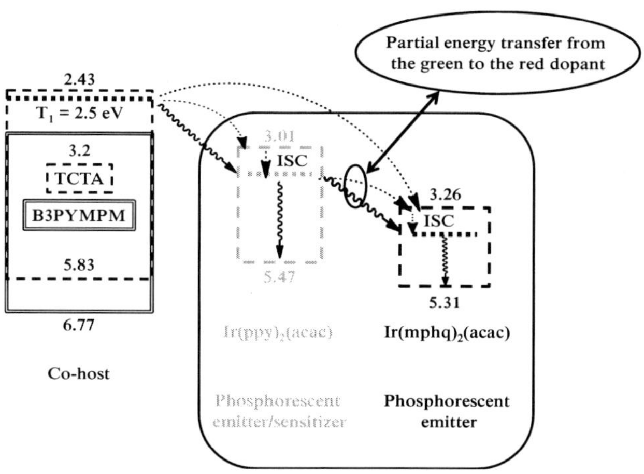

Figure 5. Working principle of a phosphor-sensitized OLED based on a partial energy transfer from the green to the red dopant.

When tested in OLEDs, an excellent color purity was obtained, and the EL spectra superimposed the PL spectra of the two complexes, evidencing a complete energy transfer from the triple co-host to the dopants. However, device performances remained quite limited, with maximum brightnesses found to be of 4024 and 7219 cd/m² for (EO-CVz-PhQ)$_2$Ir(prz) and (EO-CVz-PhQ)$_2$Ir(mprz) respectively. Higher device performance with (EO-CVz-PhQ)$_2$Ir(mprz) was assigned to the absence of energy barrier between the LUMO of the dopant and the LUMO of OXD-7, thus facilitating the charge injection and transport. Conversely, the LUMO gap of 0.15 eV between (EO-CVz-PhQ)$_2$Ir(prz) and OXD-7 required an over-potential corresponding to the energy barrier, thus reducing the device efficiency at similar driving voltage. Because of the large HOMO/LUMO level difference between (EO-CVz-PhQ)$_2$Ir(prz) and OXD-7, both holes and electrons could be trapped by the dopant, lowering the device performances. One year later, the same authors prolongated this initial study with OLEDs comprising other phenylquinoline-based complexes, namely (*bis*[9-ethyl-2-(4-phenylquinolin-2-yl)-*9H*-carbazolato-*N,C*$^{2'}$]iridium picolinate-*N*-oxide [2-Et-CVz-PhQ)$_2$Ir(pic-NO)] and *bis*[9-ethyl-3-(4-phenylquinolin-2-yl)-*9H*-carbazolato-*N,C*$^{2'}$]iridium picolinate-*N*-oxide (3-Et-CVz-PhQ)$_2$Ir(pic-NO)) [72]. To afford the electron transport ability to the dopants, carbazole moieties were introduced by varying the substitution position. To investigate the EL performances in solution-processed devices, the same three components that were used in the previous study were also mixed to form the host material, namely, OXD, PVK and TPD. As previously observed, if the solution-processed approach is appealing, performances remained however limited compared to what can be obtained with vacuum-deposited devices. A maximum brightness of only 9753 cd/m^2, a maximum CE and PE hardly reaching 8.74 cd/A and 13.86 lm/W at 30 mA/cm^2 were obtained with (3-Et-CVz-PhQ)$_2$Ir(pic-N-O) as the dopant. The roll-off remained limited to 15%, but was only examined on a limited current density range (1 to 30 mA/cm^2). Recently, the level of sophistication was increased a step further with the use of a new concept based on a cascade exciton transfer mechanism developed for the design of yellow PhOLEDs [73]. To reinforce the energy transfer from the host to the phosphorescent dopant, an organic host (TCTA) was blended with an iridium (III) complex, not used in this case as a dopant, but as a co-host for TCTA. Notably, two wide-bandgap iridium complexes were tested as effective co-hosts for TCTA, namely FIrpic and Ir(ppy)$_3$. As main advantage, metal complexes counter-balance the hole-dominated transportation of TCTA by bringing the electron-transport character, hence balancing the carrier injection in the emissive layer. To act as co-hosts, the phosphorescent guest was selected such that the host complexes exhibited suitable HOMO and LUMO levels to host the dopant (iridium(III) *bis*(4-phenylthieno[3,2-*c*]pyridinato-*N,C*$^{2'}$)acetylacetonate (PO-01)).

Using this strategy, two efficiency-effective energy transfer routes from the co-hosts to the dopant could be created: one by a direct energy transfer from TCTA to the yellow emitter and the second one, by mean of an energy transfer from TCTA to the phosphorescent co-host on which the excitons are first generated, followed in a second step by a transfer of excitons on the guest (see Figure 7). Interestingly, all the spectra of the FIrpic co-host containing devices showed no emission in the short wavelength region, indicating that an efficient energy transfer has occurred. Benefits of this "double energy transfer route" were clearly evidenced by a simple comparison. Notably, blend of FIrpic with TCTA resulted in an increase of 32% of the PE compared to that of similar devices only comprising TCTA as host. The PE (26.7 lm/W at 1000 cd/m^2) obtained with this approach constitutes the best results ever reported for solution-processed yellow OLEDs.

The same cascade mechanism was successfully applied to boost the performances of red PhOLEDs [74a]. In these devices, the tri-component EML was composed of a purely organic host (CBP), blended with an organometallic co-host (Ir(ppy)$_2$(acac)), used as an effective exciton harvester subsequently transferring its excitons on the red phosphorescent dopant *bis*(2-methyldibenzo [*f,h*]quinoxaline)(acetylacetonate)iridium Ir(MDQ)$_2$(acac) by triplet-to-triplet Dexter energy transfer. Here again, the key of the performance improvement was determined as being the energy transfer from the organometallic host to the red dopant. For this study, the following device configuration was investigated: ITO/ MoO_3 (1 nm)/ CBP (65 nm)/ CBP:Ir(ppy)$_2$(acac):Ir(MDQ)$_2$(acac) (15 nm)/ TPBI (75 nm)/ LiF (1 nm)/ Al. To prevent charge accumulation at the HTL/ETL interface, HTL and ETL materials were selected such that energy differences between HOMOs and LUMOs of the two materials were lower than 0.1 eV. At fixed dopant concentration and by slowly increasing the concentration of the organometallic host from 0 to 12 wt%, device performances proved to be higher at low host concentration as a result of the minimized self-quenching at these concentrations. 2 wt% was determined as the optimum concentration for the two complexes in CBP. CE and PE as high as 37.0 cd/A and 30.2 lm/W were achieved at 100 cd/m², corresponding to the impressive EQE of 24.8%. Energy transfer from the green emitter to the red one was clearly evidenced while examining the EL spectra, with a dominant emission peak centered at 606 nm originating from the red emitter, and an extremely small shoulder at 515 nm issued from the green emitter. Interestingly, this shoulder was detected whatever the co-host concentration was.

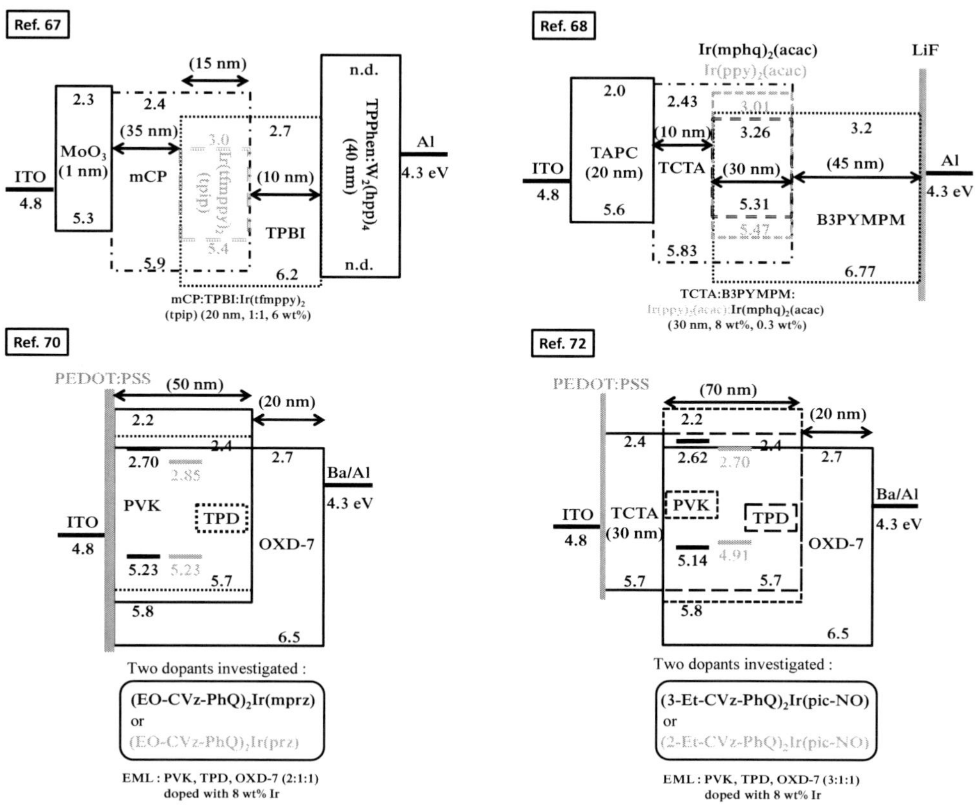

Figure 6. Energy diagrams of OLEDs designed with mixtures of co-hosts and dopants (taken from references [67, 68, 70 and 72]).

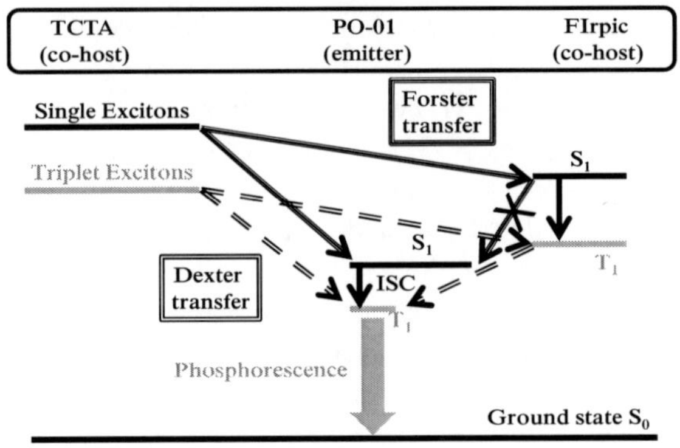

Figure 7. Excitons management in OLEDs comprising a fluorescent and phosphorescent co-host.

Meanwhile, intensity of this peak remained sufficiently low comparatively to that of the red emission such that the perceived emission color was unaffected. To get a deeper insight into this performance enhancement, the different energy transfer processes taking place in the EML were examined (see Figure 8). First, the good overlap between the CBP emission and the absorption of Ir(ppy)$_2$(acac) and Ir(MDQ)$_2$(acac) clearly demonstrated the possibility of Förster and/or Dexter energy transfer between the organic host and the two complexes. Similarly, the overlap between the Ir(ppy)$_2$(acac) emission and the Ir(MDQ)$_2$(acac) absorption was favorable to a Dexter energy transfer from the triplet state of the green dopant to the triplet state of the red one. The efficiency roll-off of devices was especially low since an EQE of 10.8% could be still detected at 5000 cd/m^2. However, limits of this double dopants strategy were recently evidenced [74b]. While using FIrpic as the co-dopant for FIr6, a saturation effect was evidenced. Notably, FIrpic's peak intensity increased gradually with the driving voltage. This process was clearly identified as resulting from a diffusion of the triplet excitons generated in FIr6 to FIrpic under high electric field, diminishing the dopant–dopant energy transfer between FIrpic and FIr6. Therefore, the radiative decay time is another parameter to consider for an efficient energy transfer from the organometallic dopant to the guest. These dual-doping systems, in which versatile desexcitons pathways are formed to channel the excitons to the dopants, were initially develop to significantly enhance the exciton harvesting efficiency on phosphors. This principle was extended to the design of white OLEDs by creating an incomplete energy transfer between the phosphorescent host and the phosphorescent dopant. As above demonstrated, emission of the phosphorescent host can be totally suppressed if the organometallic co-host is introduced at low concentration within the emissive layer. Inevitably, if its concentration increases, an incomplete Dexter energy transfer between the organometallic co-host and the dopant occurs, giving rise to the radiative desexcitation of the co-host. Using this strategy, WOLEDs were prepared by combining the emissions of FIrpic and (fbi)$_2$Ir(acac) [75]. In these WOLEDs, two main emission mechanisms were identified, resulting in turn in expending the recombination zone (see Figure 9). First, emission of FIrpic proved to be independent of its concentration, suggesting the primary emission mechanism to be based on energy transfer process. Second, direct exciton formation following charge trapping contributed to the (fbi)$_2$Ir(acac) emission and induced a significant increase of the current density with the concentration. Jointly, electron transportation by mean of (fbi)$_2$Ir(acac) in the hole-dominated host was favored by the perfect alignment of its LUMO level with those of the adjacent ETL, broadening the recombination zone and reducing the concentration quenching. This principle of incomplete energy transfer was revisited a few years later by the same authors with the same iridium complexes, still to design WOLEDs, but in a slightly different context [76]. Notably, the organic host previously used (CBP) was a unipolar host but, this time, a bipolar tetraarylsilane integrating an electron-donating arylamine and an electron-accepting benzimidazole was employed. Using (4-{diphenyl[4-(1-phenyl-1*H*-benzimidazol-2-yl)phenyl]silyl}phenyl)diphenylamine (*p*-BISi-TPA) as host, WOLEDs with a maximum brightness of 45398 cd/m², a maximum CE and PE of 51.8 cd/A and 42.7 lm/W were respectively obtained. To observe the double emission, concentration of the orange dopant $(fbi)_2Ir(acac)$ was extremely low (0.67 wt%) compared to the concentration of FIrpic (7 wt%).

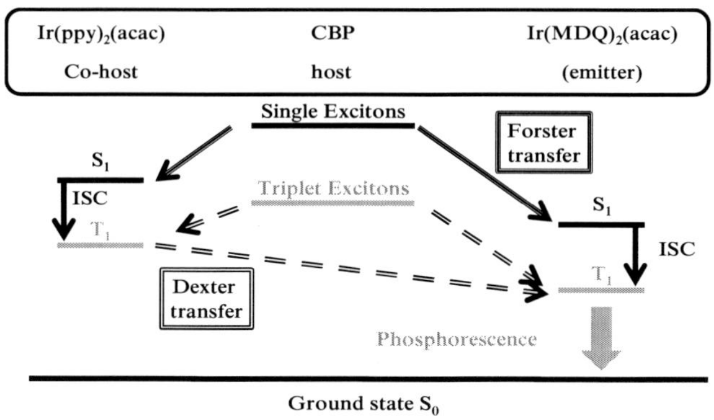

Figure 8. Excitons management in OLEDs comprising a fluorescent and phosphorescent co-host.

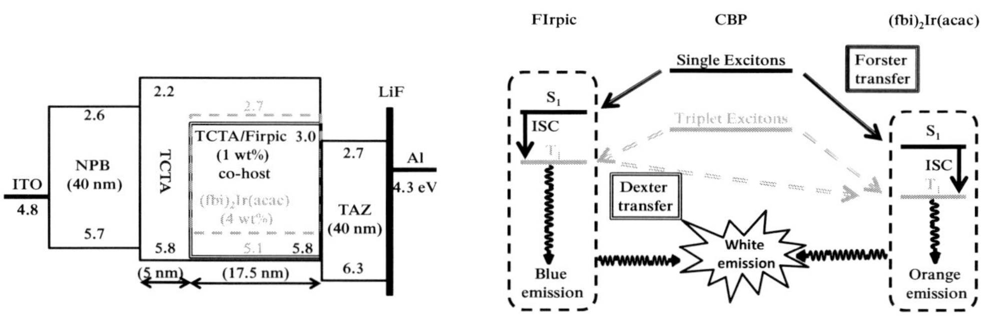

Figure 9. White OLEDs based on the incomplete energy transfer from the organometallic co-host (FIrpic) to the orange triplet emitter ($(fbi)_2Ir(acac)$).

Compared to the reference devices fabricated with mCP that exhibited strong efficiency roll-off, the low roll-off observed with *p*-BISiTPA was assigned to the bipolar nature of the host, capable of conducing equally holes and electrons within the EML, resulting in a broadening of the recombination zone and a reduction of the concentration quenching and, in turn, in higher performances. More recently, novel bipolar tetraarylsilane-based hosts furnished slightly lower performances in double-emission-layer WOLEDs as the maximum CE and PE only reached 41.5 cd/A and 39.4 lm/W respectively [77]. To further compliment the recent progress made in terms of bipolar hosts for efficient deep blue OLEDs, investigations on co-hosted devices were not limited to iridium complexes and cyclometalated Pt complexes were also studied as potential light-emitting candidates [78]. A series of four platinum complexes was notably synthesized as light-emitting materials. For this study, a blend of 2,8-*bis*(diphenylphosphoryl)dibenzothiophene (PO15) and TAPC was selected as a potential host capable to replace 2,6-*bis*(*N*-carbazolyl)pyridine (26mCPy) [79]. Using PO15:TAPC as mixed layer and Pt-16 as dopant (2 wt%), a maximum PE of 19.3 lm/W could be obtained, far from 8.9 lm/W obtained with 26mCPy in a similar device setting.

Importantly, PE obtained with PO15/TAPC as co-host were comparable to the best results reported for iridium-based devices [80]. Another approach that can also be classified among the co-host strategy consists in fabricating devices with crossfading-hosts (see Figure 10) [81].

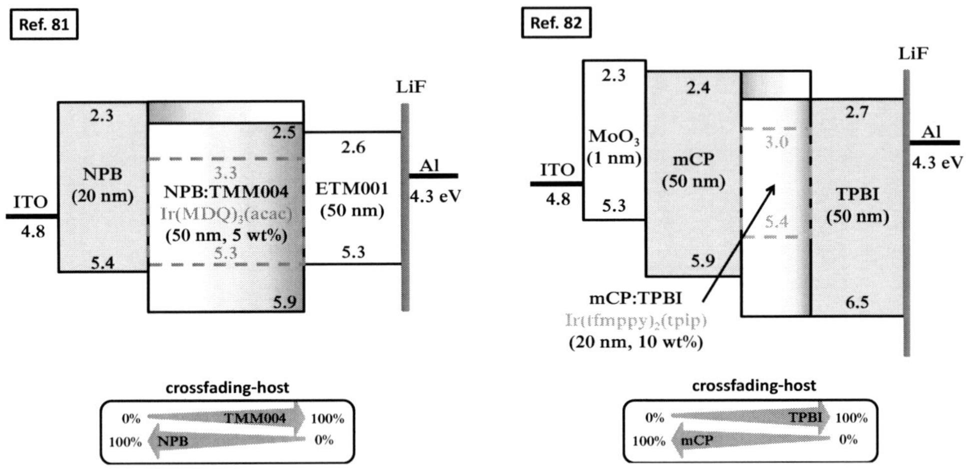

Figure 10. Energy diagram of OLEDs fabricated with cross-fading hosts and reported in references 81 and 82.

This strategy is only scarcely used as it involves the consumption of many materials and severely complicates the device fabrication. However, it can diminish the charge carrier accumulation at the interface between the emission layer and the transporting layers by mixing in the EML, the materials also used for the adjacent HTL and ETL layers. All energy barriers with the adjacent layers are thus suppressed. Red PhOLEDs reaching the high PE of 33.6 lm/W was achieved at the luminance of 1000 cd/m^2 and the roll-off in the 1 000-10 000 cd/m^2 range could be neglected. As a logical extension of this work, a comparison between double-hosts, co-hosts and crossfading hosts-based devices evidenced the superiority of the last approach to address the roll-off issue [82]. As main benefit of the gradient of materials, it was found out that the recombination zone could expand with increasing the current density. This phenomenon was easily demonstrated by introducing sensing layers on both sides of the EML and the observation of their respective emissions only at high current density. Modularity of the width of the recombination zone with the operating voltage is a remarkable tool for reducing the concentration quenching and improving device performances.

3. Broadening the Recombination Zone with Multi-Emission-Layer Devices

The previous paragraph has highlighted the benefits of mixing a hole and an electron transport material for balancing the charge transportation within the EML. Still with the idea of combining two well-known hole and electron transport materials for the design of EMLs, a comparative study was carried out between the performances of double and single-layered EML devices for the design of orange PhOLEDs. If the work of Zheng and co-workers [82] evidenced the higher performances of single EML devices fabricated by blending two co-hosts, the opposite conclusions were deduced from the present work that used TCTA and TPBI as host materials for *tris*(2-phenylquinoline)iridium (III) Ir(2-phq)$_3$ (see Figure 11) [83]. Effectiveness of the double emissive EML strategy was outstandingly demonstrated since the record-breaking PE of 47.2 lm/W at 100 cd/m² was obtained in 2010 for these orange PhOLEDs. Similarly, EQE only decreased of 8% in the 100-1 000 cd/m² range. Conversely, single EML devices comprising a mixed host EML in the 7:3, 1:1 and 3:7 ratios only reached 24.8, 21.1 and 18.8 lm/W respectively at the same brightness while exhibiting the same roll-off than double EML devices. This strategy of double EML is not new and was previously investigated with other combinations of materials (BPhen and TCTA used as hosts for Ir(ppy)$_3$, [84] mCP and UGH3 used as hosts for FIrpic), [85] but without significant control on the roll-off.

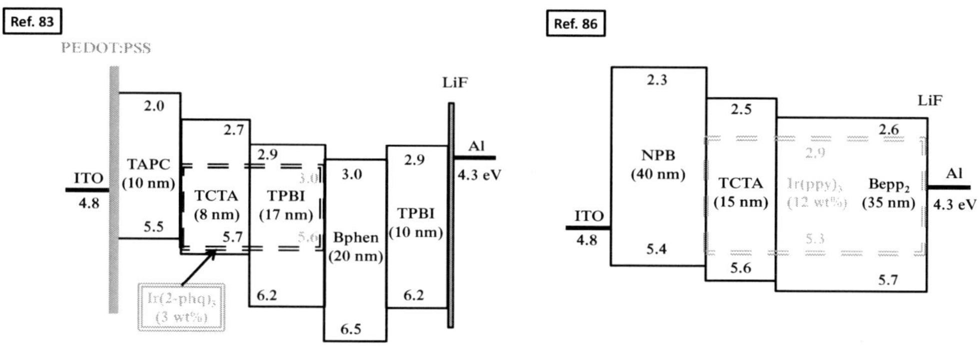

Figure 11. Structure of OLEDs investigated to compare single and double emissive layers.

Double EMLs devices associating an organic and an organometallic host were also investigated as part of an effective alternative to purely organic-based devices. The premise of this strategy was reported five years ago [86]. In this approach, the two hosts, TCTA and Bepp$_2$, were simply deposited on top of each other and doped with Ir(ppy)$_3$ at 12 wt%. Especially, TCTA was selected for its excellent hole-transport mobility and Bepp$_2$ for its electron-transport mobility close to 10^{-4} cm²/Vs [87]. Additionnally, evolution of their respective HOMO and LUMO levels enabled to form a stepped progression of their energy levels, favorable to a reduction of the driving voltage. Device structure could also be profoundly simplified since the charge transport layers also acted as co-emissive layers. Efficacy of the double host-layer strategy was clearly evidenced with an EQE variation lower than 11% in the 1 000 to 10 000 cd/m² region. Similarly PE and CE as high as 65.1 lm/W and 58.7 cd/A only rolled off to 53.9 lm/W and 57.5 cd/A at 1000 cd/m², corresponding to a roll-off of only 6% for the CE. Enhancement of the device performances, relative to that obtained with the single layered-devices only comprising $Bepp_2$ or TCTA as host, was notably assigned to a higher confinement of excitons inside the light-emitting layer and a better balanced hole and electron injection and transportation. Improvement of the device performance also originates from exciton formation by carrier recombination arising at the interface of TCTA:$Ir(ppy)_3$/$Bepp_2$:$Ir(ppy)_3$ and a recombination zone sufficiently away from the NPB/TCTA interface and cathode, thereby preventing any quenching of radiative excitons. By changing the dopant concentration from 12 to 15%, slightly lower performances were obtained. Interestingly, the triplet-triplet annihilation could be totally suppressed and no variation of the CE was observed in the 0-1 000 cd/m² range. The strategy, that employs a double EML, has clearly demonstrated its high efficiency and a reduced efficiency roll-off compared to conventional single-emission-layer OLEDs due to an expanded excitons formation zone. Efforts were thus devoted to transpose this approach to WOLEDs. Notably, WOLEDs were fabricated by stacking the three complementary red-green-blue emitters into three adjacent layers, each layer comprising its own host material doped with one of the RGB phosphorescent emitter (see Figure 12) [88]. The three host materials were arranged such that a stepped progression of HOMO levels was observed (from TCTA (5.7 eV) to mCP (5.9 eV) to UGH2 (7.2 eV)). However, the large energy barrier existing between the HOMOs of mCP and UGH2 could result in holes accumulation at this interface. To avoid this, UGH2 was doped at 20 wt%, therefore promoting the charge injection directly on the dopant. High efficiency of WOLEDs was assigned to the convenient choice of both hosts and dopants in the adjacent layers such that the HOMOs and LUMOs of each host was aligned with the HOMOs and LUMOs of the dopants in the adjacent emissive layers, resulting in a broad recombination zone. A non-negligible roll-off of the PE was observed, since the maximum PE of 64 ± 3 lm/W was reduced to 34 ± 2 lm/W at 1000 cd/m², corresponding to a decrease of 47%. Still based on RGB emitters located into three adjacent EMLs, the roll-off issue could be addressed by using a blend of co-host for two of the three EMLs (see Table 2) [89]. By mixing TCTA and TPBI in a 1:3 ratio for the two EMLs adjacent to the HTL and the ETL, a higher balance of the hole and electron transportation was achieved, thus reducing the roll-off. To ensure that the recombination zone extends to the blue central emissive layer, very thin red and green EMLs were utilized to favor the charge transport of holes and electron.

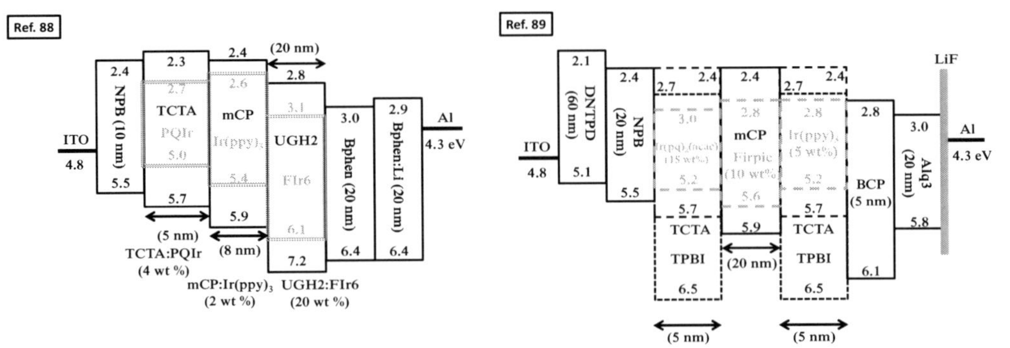

Figure 12. WOLEDs fabricated by introducing the three complementary RGB colors in adjacent layers (taken from references 88 and 89).

Table 1. Device Characteristics of OLEDs addressing the roll-off issue

Host	Device structure	V_{ON} (V) [a]	Max η_c (cd/A)	Max η_p (lm/W)	CIE coord.	Max brightness (cd/m^2)	EQE (%)	λ_{EL} (nm)	T_g (°C)	Ref.
B3PYMPM	ITO(100 nm)/TAPC(30 nm)/ TCTA(10 nm)/ TCTA:B3PYMPM: $Ir(ppy)_2(acac)$(30 nm)/ B3PYMPM(40 nm)/ LiF(0.7 nm)/ Al	2.4		124.0			29.1			[57]
SPPO13	ITO/ DNTPD(60 nm)/ NPB(20 nm)/ TAPC(10nm)/ TAPC:SPPO13: FIrpic (30 nm, 10%)/Bphen(20 nm)/ LiF(1 nm)/Al			28.4 at 1 000 cd/m^2		10 000 cd/m^2 at 7.8V	17.9% at 1 000 cd/m^2			[58]
TpPyPB	ITO/TAPC(10 nm)/TAPC: TpPyPB: $Ir(ppy)_3$(1:1, 5%, 70 nm)/LiF(0.5 nm)/ Al	2.0	32.8	29.1	0.31, 0.63		13.8	515		[59]
SPPO13	ITO/PEI:Li_2CO_3/ SPPO13: TCTA:FIrpic(60 nm) TCTA (40 nm)/ MoO_3(5 nm)/Al	4.3	34.3	20.5	0.16, 0.30		15.2			[60]
SPPO13	ITO/PEI:Li_2CO_3/SPPO13:TCTA:$Ir(mppy)_3$(60 nm) TCTA (40 nm)/MoO_3(5 nm)/Al	3.9	56.9	35.4	0.32, 0.61		16.3			[60]
SPPO13	ITO/PEI:Li_2CO_3/SPPO13:TCTA: $Ir(MQD)_2(acac)$ (60 nm) TCTA(40 nm)/MoO_3(5 nm)/ Al	4.5	12.0	7.9	0.65, 0.35		10.8			[60]
UGH3	ITO/TAPC(50nm)/TCTA: UGH3: FIrpic(30 nm)/ TmPyPB (20nm)/ TmPyPB:Cs (30nm)/LiF(1)nm)/Al	3.0		44.9			21.6			[63]
UGH3	ITO/TAPC(50 nm)/ TCTA: UGH3: Fir6(30 nm)/TmPyPB(20nm)/ TmPyPB:Cs (30nm)/LiF(1)nm)/Al	3.0	18.7	15.0			10.9			[64]
SPPO1	ITO(70nm)/TAPC(50 nm)/ TCTA: SPPO1:FIrpic(30 nm)/ Tm3PyPB(20 nm)/Tm3PyPB:Cs(30 nm)/LiF(1 nm)/Al	2.5		53			21.8			[65]
TmPyPb	ITO)/MoO_3(3 nm)/TCTA (50 nm)/TCTA:TmPyPb:FIrpic (20 nm)/TmPyPb (30 nm)/ LiF(1 nm)/Al	2.68		54.5			20.4			[66]
TCTA: TPBI	ITO/PEDOT:PSS(35 nm)/ TAPC (10 nm)/TCTA:TPBI: $Ir(2\text{-}phq)_3$ (25 nm)/BPhen (20 nm)/TPBI (10 nm)/ LiF (0.7 nm)/Al	3.0	47.4 at 100 cd/m^2	47.2 at 100 cd/m^2			20.3			[83]
mCP:TPBI	ITO/MoO_3(1 nm)/mCP (60nm)/ mCP:TPBI: $Ir(tfmppy)_2(tpip)$ (20 nm)/ TPBI (15 nm)/ TPPhen: $W_2(hpp)_4$(45 nm, 12 wt%)/ Al		72.9	66.3			20.8			[67]
TCTA: B3PYMP: $Ir(ppy)_2(acac)$	ITO/ TAPC (20 nm)/ TCTA (10 nm)/ TCTA:B3PYMPM: $Ir(ppy)_2(acac)$: $Ir(mphq)_2(acac)$ (30 nm, 8 wt%, 0.3 wt%)/ B3PYMPM (45 nm)/ LiF (0.7 nm)/ Al	2.4		62.1	0.56, 0.42	11730 at 30 mA/cm^2	25.0			[68]
PVK:TPD:OXD-7	ITO/PEDOT:PSS/TCTA (30 nm)/PVK: TPD:OXD-7: (EO-CVz-PhQ)2Ir(prz) (50 nm)/OXD-7 (20 nm)/Ba (3 nm)/Al	5.6	6.69	1.80	0.60, 0.39	7219	3.68	584		[70]

Table 1. (Continued)

Host	Device structure	V_{ON} (V) [a]	Max η_c (cd/A)	Max η_p (lm/W)	CIE coord.	Max brightness (cd/m²)	EQE (%)	λ_{EL} (nm)	T_g (°C)	Ref.
PVK:TPD:OXD-7	ITO/PEDOT:PSS/TCTA (30 nm)/PVK: TPD:OXD-7:(EO-CVz-PhQ)$_2$Ir(mprz) (50 nm)/OXD-7 (20 nm)/Ba (3 nm)/Al	5.6	3.96	1.03	0.60, 0.39	4024	2.33	582		[70]
TCTA: FIrpic	ITO/PEDOT:PSS/TCTA:FIrpic(7.5wt%):PO-01 (15wt%)/TPBI (32 nm)/LiF (0.7 nm)/Al		47.6 at 100 cd/m²	35.1 at 100 cd/m²	0.50, 0.50	32260	14.6 at 100 cd/m²			[73]
CBP:Ir(ppy)$_2$(acac)	ITO/MoO_3(1nm)/CBP(65nm)/CBP:Ir(ppy)$_2$(acac):Ir(MDQ)$_2$(acac)(15nm)/ TPBI(75nm)/LiF:Al	2.8	37				24.8	606		[74]
TCTA (5 nm)/ *p*-BISiTPA:FIrpic	ITO/MoO_3 (10 nm)/NPB (80 nm)/TCTA (5 nm)/ *p*-BISiTPA: 8 wt% FIrpic: 0.67 wt% (fbi)$_2$ Ir(acac) (20 nm)/TPBI (40 nm)/LiF (1 nm)/Al		51.8	42.7	0.38, 0.44	45 398	19.1			[76]
DCzSiPy	ITO/MoO_3(2 nm)/DDCzSi(40 nm)/DCzSiPy : FIrpic(8 wt%) (10 nm) /DCzSiPy : PO-01(6 wt%) (10 nm)/Bphen(40 nm)/LiF(0.8 nm)/Al		41.5	39.4	0.42, 0.47		13.7			[77]
TAPC:PO15	ITO/PEDOT:PSS/ NPD (30 nm)/ TAPC (10 nm)/2% Pt-16:TAPC: PO15(25 nm)/ PO15(40 nm)/ LiF/ Al			23.3	0.16, 0.13		14.8			[78]
TCTA-bepp$_2$ bilayer	ITO/NPB (40 nm)/ TCTA:Ir(ppy)$_3$ (15 nm, 15 wt%) /Bepp$_2$:Ir(ppy)$_3$ (35 nm, 15wt %) /LiF (1 nm)/ Al	2.5	51.2	57.9	0.32, 0.61		16.6			[86]

Table 2. Device performances of OLEDs addressing the roll-off issue

Host	Device structure	V_{ON} (V) [a]	Max η_c (cd/A)	Max η_p (lm/W)	CIE coord.	Max brightness (cd/m²)	EQE (%)	λ_{EL} (nm)	Ref.
TCTA-mCP-UGH2	ITO/ NPD (40 nm)/ 4 wt % PQIr:TCTA (5 nm)/ 2 wt% $Ir(ppy)_3$:mCP (8 nm)/ 20 wt % FIr6:UGH2 (20 nm)/ Bphen (20 nm)/ Bphen:Li (20 nm)/ Al			64	0.37, 0.41		26		[88]
TCTA: TPBI/ MCP/ TCTA: TPBI	ITO/ DNTPD (60 nm)/ NPD (20 nm)/ TCTA (10 nm)/ TCTA:TPBI: $Ir(pq)_2(acac)$ (5 nm,15 wt%)/ mCP:Firpic (20 nm, 10 wt%)/ TCTA:TPBI:$Ir(ppy)_3$ (5 nm, 5wt%)/ BCP (5 nm)/ Alq3 (20 nm)/ LiF (1 nm)/ Al						13 at 1000 cd/m²		[89]
PVK: OXD-7:TPD	ITO/ PEDOT:PSS (40 nm)/ TCTA(30 nm)/ PVK:OXD-7:TPD: $(3\text{-Et-CVz-PhQ})_2$Ir(pic-N-O) (70 nm)/ OXD-7 (20 nm)/ Ba (3 nm)/ Al		13.9	4.32	0.63, 0.37	9753	8.7	595	[72]
PVK: OXD-7:TPD	ITO/PEDOT:PSS (40 nm)/ TCTA (30 nm)/ PVK: OXD-7:TPD:$(2\text{-Et-CVz-PhQ})_2$Ir(pic-NO) (70 nm)/ OXD-7 (20 nm)/ Ba (3 nm)/ Al		0.5	0.16	0.68, 0.28	618	4.3	668	[72]
TCTA	MeOTPD:4F-TCNQ (4 wt%, 25 nm)/ NPB (10 nm)/ [TCTA (2 nm)/ TCTA:$Ir(ppy)_3$(1 nm)]$_6$ (20 nm)/ TPBI(10 nm)/Bphen:Cs (55 nm)/ Al	2.3		36.5		4800 at 10 mA/cm²	14.4		[92]
	ITO/ NPB (45 nm)/ NPB:DCJTB (10 nm, 0.5 wt.% DCJTB)/ CBP (10 nm)/ CBP: $Ir(piq)_3$ (20 nm, 7 wt.%)/ TPBI (40 nm)/ LiF (1 nm)/ Al		6.0						[93]
	ITO/ NPB (45 nm)/ NPB:C545T(10 nm, x wt %)/ CBP(10 nm)/ CBP:Ir(H-$Cz\text{-}py)_3$(20 nm, 7 wt %)/ TPBI(40 nm)/ LiF(1 nm)/ Al	9.0	18.5						[95]
	ITO/NPB (45 nm)/ NPB:DSA-Ph (10 nm, 3 wt.% DSA-Ph)/ CBP (10nm)/ CBP:FIrpic (20 nm, 7wt.% FIrpic)/ TPBI (40nm)/ LiF (1 nm)/ Al	5.5	15.8	10.2	0.23, 0.47	13085			[96]
	ITO/Ni-100 (50 nm)/ BPAPF (15 nm)/ BH-15:EB-502 (2%, 8 nm)/BH-05 (2 nm)/ EPH-31 (2 nm)/ EPH-31:Os (2%, 26 nm)/ EPH-31 (2 nm)/ CsF (2 nm)/Al	6.3	15.2	7.8	0.31, 0.37	2800			[97]
	ITO/MoO_3(8 nm)/TCTA(60 nm)/TCTA:$Bepp_2$ (50:50): $Ir(ppy)_2(acac)$(6%, 5nm)/ TCTA: $Bepp_2$(50:50):$Ir(MDQ)_2(acac)$ (6%, 5 nm)/ TCTA:$Bepp_2$ (100:15, 3 nm)/ $Bepp_2$:DSA-ph(2.5%, 10 nm)/ $Bepp_2$(30 nm)/LiF (1 nm)/Al		30.8	27.1	0.43, 0.43	1000 cd at 3.56V	13.7		[98]
	ITO/ MCC-PC 1020 (20 nm)/ 3DTAPBP (20 nm)/ TCTA:FIrpic (5 nm)/TCTA: PQ_2Ir (3 wt%, 0.25 nm)/ DCzPPy:PQ_2Ir (3 wt%, 0.25 nm)/ DCzPPy):FIrpic (5 nm)/ BmPyPB (50 nm)/ LiF (0.5 nm)/ Al			53 at 100 cd/m²			25% at 1000 cd/m²		[101 a]
	ITO/ (2T-NATA (50 nm)/ NPB (5 nm)/ TCTA (8 nm)/ mCP:2 wt% Pt-4 (8 nm)/ TPBI: 5 wt% PQIr (0.25 nm)/ TPBi (0.5 nm)/ TPBI: 6 wt% $Ir(ppy)_3$ (0.75 nm)/ TPBI (30 nm)/ LiF (0.3 nm)/ Al		29.2						[102]

Table 2. (Continued)

Host	Device structure	V_{ON} (V)[a]	Max η_c (cd/A)	Max η_p (lm/W)	CIE coord.	Max brightness (cd/m²)	EQE (%)	λ_{EL} (nm)	Ref.
	ITO/ MeO-TPD:F_4-TCNQ (4 wt%, 60 nm)/ TAPC (10 nm)/ 4P-NPD: $Ir(MDQ)_2$(acac) (0.2 wt%, 10 nm)/ TCTA:TPBI (1:3, 3 nm)/ TPBI (10 nm)/ BPen:Cs (30 nm)/ Al			40.7 at 1000 cd/m²	0.43, 0.43		20.3% at 1000 cd/m²		[103]
	ITO/ 2-TNATA (5 nm)/ NPB) (40 nm)/ CBP:$Ir(ppy)_3$ (5 wt%): DMQA (0.5 wt%) (30 nm)/ Bphen (10 nm)/ Alq_3 (20nm)/ LiF (0.5 nm)/ Al						5.35		[104]
	ITO/ TAPC (40 nm)/ 1 or 2 (x wt%):mCP (20 nm)/ TPBI (40 nm)/ LiF (1 nm)/Al								[105]
	ITO)/ NPB (50 nm)/ $(tbpbt)_2Ir$(acac):CBP (30 nm)/ BCP (10 nm)/ Alq_3 (40 nm)/Mg:Ag	5.4	26.9	8.41	0.56, 0.43	29 800 at 18.9 V	9.44	580	[108]
	ITO)/ NPB (50 nm)/ $(fbt)_2Ir$(acac):CBP (30 nm)/ BCP (10 nm)/ Alq_3 (40 nm)/Mg:Ag	5.3	28.5	10.85	0.63, 0.36	25 900 at 20 V	15.6	600	[108]
	ITO/NPB (50 nm)/$(t\text{-}bt)_2Ir$(acac) (1 nm)/NPB (0 nm)/ 2,9-dimethyl-4,7-diphenyl-phenanthroline (BCP) (30 nm)/Mg:Ag 10:1 (100 nm)			11.04	0.502, 0.492				[114]
	ITO/NPB (50 nm)/$(t\text{-}bt)_2Ir$(acac) (1 nm)/NPB (6 nm)/ 2,9-dimethyl-4,7-diphenyl-phenanthroline (BCP) (30 nm)/Mg:Ag 10:1 (100 nm)			20.26	0.485, 0.451				[114]
	ITO/ NPB (30 nm)/ TAPC (20 nm)/ CBP (15 nm)/ $(pbi)_2Ir$(acac) (x nm)/ CBP (15nm)/ BPhen (40 nm)/ Mg:Ag (200 nm)	5.1	36.2	32.4	0.33,0 .42	26 531		594	[116]
	ITO/PEDOT:PSS (50 nm)/Ir-G1 (80 nm)/TPBI (25 nm)/Ba (4 nm)/Al	2.8	40.9 at 1000 cd/m²	39.5 at 1000 cd/m²	0.50, 0.48		16.4 at 1000 cd/m²	571	[117]
	ITO/PEDOT:PSS (50 nm)/Ir-G2 (80 nm)/TPBI (25 nm)/Ba (4 nm)/Al	3.0	38.9 at 1000 cd/m²	38.3 at 1000 cd/m²	0.52, 0.47		15.6 at 1000 cd/m²	568	[117]
$Be(PPI)_2$	ITO/ MoO_3(10 nm)/ NPB (60 nm)/ TCTA(5 nm)/ $Be(PPI)_2$:$Ir(ppy)_3$ (30 nm)/ TPBI (30 nm)/LiF (1 nm)/Al	2.7	55.6	61.4	0.32,0 .61	92 261	15.3		[118]
TPBI: $Ir(ppy)_3$	ITO/ PEDOT:PSS (25 nm)/ TAPC (20 nm)/ TPBI: $Ir(ppy)_3$ (10 wt%):PO-01 (5.8 wt%) (20 nm)/TPBI (32 nm)/ LiF (0.7 nm)/ Al		71 at 1000 cd/m²	59 at 1000 cd/m²			23 at 1000 cd/m²		[119]

Host	Device structure	V_{ON} (V) [a]	Max η_c (cd/A)	Max η_p (lm/W)	CIE coord.	Max brightness (cd/m^2)	EQE (%)	λ_{EL} (nm)	Ref.
	ITO/ NPB(50 nm)/ CBP: BCzVBi (10 wt%, 10 nm)/ CBP(2 nm)/ CBP: $Ir(ppy)_3$ (8 wt%): DCJTB (0.08 wt%) (15 nm)/ CBP (2 nm)/ BCP:BCzVBi (10 wt%, 10 nm)/ BPhen (30 nm)/ LiF (0.8 nm)/Al		22	18.1			8.5		[120]
	ITO/ MoO_3 (10 nm)/ NPB (60 nm)/ TCTA (5 nm)/ $Zn(PPI)_2$:$Ir(ppy)_3$ (30 nm)/ TPBI (30 nm)/LiF (1 nm)/Al	2.7	58.0	67.5	0.30,0.63	85 872	15.9		[118]
	ITO/ MoO_3 (10 nm)/ NPB (60 nm)/ TCTA (5 nm)/ $Be(PPI)_2$: $Ir(MDQ)_2$(acac) (30 nm)/TPBI (30 nm)/LiF (1 nm)/Al	2.3	15.9	19.9	0.64,0.36	38 580	15.1		[118]
	ITO/ MoO_3 (10 nm)/ NPB (60 nm)/ TCTA (5 nm)/ $Zn(PPI)_2$: $Ir(MDQ)_2$(acac) (30 nm)/TPBI (30 nm)/LiF (1 nm)/Al	2.3	15.9	21.7	0.64,0.36	39 138	15.2		[118]
	ITO/ MeO-TPD:F_4-TCNQ (60 nm)/ NPD(10 nm)/NPD: $Ir(MDQ)_2$(acac) (20 nm)/ 4P-NPD (15 nm)/ TPBI:$Ir(ppy)_3$(10 nm)/ BPhen(10 nm)/BPhen:Cs (30 nm)/ Al	2.1		22.0		1000 cd/m^2 at 3.46V	10.4		[122a]
	ITO/ PEDOT:PSS (30 nm)/ PVK:$Ir(ppy)_3$:DCJTB (100:5:0.4 in wt.) (60 nm)/ NPB (4 nm)/ BCP (10 nm)/ Alq3 (20 nm)/ LiF (1 nm)/Al			4	0.25, 0.27	8700	6.9		[123]
	ITO/ PEDOT:PSS (30 nm)/ NPB (100 nm)/ CBP:Ir(Cz-CF_3) (25 nm)/ NPB (3 nm)/ MADN:DSA-Ph (40 nm)/ TAZ (10 nm)/ LiF (1 nm)/Al		42.5 at 1000 cd/m^2	24.9 at 1000 cd/m^2					[110g]
	glass/ PEDOT:PSS (40 nm)/ Al (50 nm)/ LiF (2.5 nm)/ TpPyPB (40 nm)/ CBP: $Ir(ppy)_3$ (20 nm)/ TAPC (35 nm)/ HAT-CN (5 nm)/ Ag (20 nm)/ NPB (80 nm)	3.0	119 cd/m^2 at 100 cd/m^2						[126]

As the driving voltage increased, a decrease of the red emission concomitantly with the increase of the blue emission was observed in the EL spectra, evidencing a clear shift and a broadening of the recombination zone at high voltage.

The maximum EQE was 14.4% and the quantum efficiency at 1000 cd/m^2 was 13.0%. Generating a broadband white light typically requires two or three emitters. The strategy based on three emitters has been detailed just above. Double-emissive WOLEDs based on a dichromatic blue/green emission is a viable alternative to triple-doped OLEDs [90]. To produce a stable white emission, an efficient carrier confinement should be obtained and the condition to reach it is to use a specific host for each EML that varies from the host in the adjacent layer by its own hole and electron-transport ability. This strategy was developed with 2,2'-*bis*[5-phenyl-2-(1,3,4)-oxadazolyl]biphenyl (OXD) as host for Ir(ppy)$_3$ and mCP as host for FIrpic. By fine tuning the dopant concentrations in the two EMLs, carrier confinement was automatically obtained, enlarging the recombination zone and stabilizing the EL emission. Lastly, outstanding CE and PE of 35.8 cd/A and 20.5 lm/W were respectively achieved with a variation of the CIE coordinates as low as 0.014 and 0.004 in the 100-10 000 cd/m^2 range. Recently, single-emitter WOLEDs have been developed by splitting the emitter into three adjacent emissive layers [91]. By varying the concentration of the platinum dopant in two of the three emissive layers, by using a neat layer of the emitter and by modifying the relative position of the different layers, WOLEDs could be obtained. EQE retention higher than 90% at 1000 cd/m^2 was obtained with *bis* [3,5-*bis*(2-pyridyl)-1,2,4-triazolato]platinum (II) [Pt(ptp)$_2$] as the emitter. For the optimized structure, peak PE as high as 30.4 lm/W, EQE of 17.1%, CE of 47.5 cd/A were achieved. Only a decrease of 14% could be observed for the EQE. If this original strategy has only been developed with platinum complexes, it could easily been transposed to iridium complexes.

4. Devices with Fluorescence-Interlayer-Phosphorescence (FIP EML) Emissive Structure

To address the roll-off issue, other sophisticated approaches were also envisioned. As first example, an EML formed by a blend of Ir(ppy)$_3$ and TCTA was fragmented into seven sub-EMLs by intercalating six neat layers of TCTA between each doped TCTA layer (see Figure 13) [92]. Using this strategy, TCTA interlayers efficiently acted as exciton blocking layers, confining excitons within each sub-EML and limiting the triplet exciton diffusion. TTA was suppressed and the roll-off behavior was greatly improved. Another fruitful design tactic was also recently proposed. As main difference over the previous strategies, the emissive layer was constituted of three stacked layers, one doped with a fluorescent emitter, one doped with a phosphorescent emitter, these two layers containing a different host and being separated from each other by an interlayer [93]. Using this fluorescence-interlayer-phosphorescence emission layer structure (FIP EML), efficiency roll-off of red PhOLEDs was reduced to 39% in the 10-500 mA/cm^2 range, contrarily to the conventional single EML devices that exhibit roll-off close to 60%. For this study, 4-(dicyanomethylene)-2-*tert*-butyl-6-(1,1,7,7-tetramethyljulolidyl-9-enyl)-4*H*-pyran (DCJTB) was selected as the fluorescent red dopant, and Ir(piq)$_3$ as the phosphorescent red emitter. For comparison and to evidence the benefit of this strategy, reference devices only bearing DCJTB or Ir(piq)$_3$ as dopants were elaborated. As anticipated, devices built up with DCJTB as unique dopant proved to be four times less emissive than the Ir(piq)$_3$-devices, demonstrating that all singlets and triplets excitons of the phosphorescent dopant are effectively harvested for light emission. A severe efficiency roll-off reaching 60% was observed in the 10–500 mA/cm^2, as classically observed for PhOLEDs. Remarkably, the FIP approach significantly limited the efficiency roll-off that was reduced to only 20% over the same current density range. Presence of the CBP interlayer was also determined as being compulsory. Indeed, elimination of the CBP layer was detrimental for the roll-off that plummets to zero on an extremely narrow current density range. Especially, application of high current densities could even not be envisioned for these devices, breakdown even occurring at low current density. Reduction of the roll-off was notably assigned to the specific operating principle of these FIP devices (see Figure 13). First, excessive hole-injection in the CBP:Ir(piq)$_3$ could be efficiently modulated, and in this aim, a fraction of the hole carriers was trapped by DCJTB whereas electrons were efficiently confined within the CBP:Ir(piq)$_3$ layer due to the presence of the CBP electron blocking interlayer.

Balance of holes and electrons is thus improved. Simultaneously, part of the excitons of the CBP:Ir(piq)$_3$ layer diffuse inside the CBP interlayer, reducing the excitons density and thus avoiding the negative effect of the triplet-triplet annihilation. CBP interlayer thickness was higher than 10 nm, corresponding to the triplet diffusion length [94]. Presence of polarons is also reduced in the CBP and CBP:Ir(piq)$_3$ layers by the reduction of the hole current, therefore jointly reducing the triplet-polaron quenching. In turn, the overall device performance is improved.

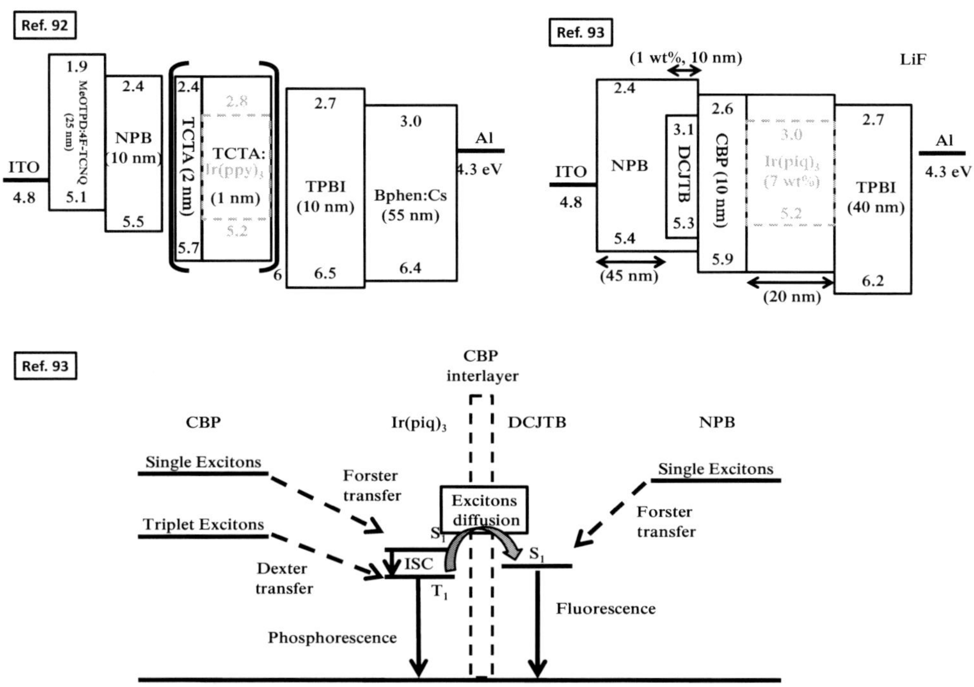

Figure 13. Devices with FIP emissive layers (taken from reference [92] and [93]).

Based on the same strategy, the same authors also fabricated green PhOLEDs [95]. Comparison with single layer devices proved once again the FIP EML structure to lower the efficiency roll-off to 26% whereas conventional phosphorescent OLEDs with single EML showed a roll-off close to 51% when operated in the same 5–150 mA/cm^2 range. Authors having developed this strategy for red and green OLEDs, finally, blue-emitting devices were fabricated as well [96]. In this work, a blue fluorescent and a blue phosphorescent emitter were jointly used for light generation. Especially, the fluorescent blue emitter was utilized to accelerate the exciton generation on the blue phosphor, the non-emissive triplet excitons of the fluorescent emitter and the triplet excitons of the phosphorescent emitter being both harvested for light emission by the blue phosphor, thereby sustaining the efficiency at current density. The second key-role played by the fluorescent dopant also consisted in favoring the hole transportation in the phosphorescent EML, reducing electron transport channels and accelerating the generation rate of excitons. Reduction of the efficiency roll-off at high luminous was thus obtained. The interlayer strategy was also fruitfully exploited for the control of the recombination zone (ZR) of hybrid white organic light emitting diodes (HWOLEDs) with fluorescence and phosphorescence hybrid structures [97]. To achieve the white emission, the blue fluorophor and the red phosphor were inserted into two distinct emissive layers separated by an interlayer. Here again, the thickness of the interlayer was careful controlled such that singlet energy transfer from the wide bandgap dopant to the low bandgap dopant was prevented. Optimum thickness was determined as being 4 nm. In this case, possibility of triplet diffusion (10 nm) was not considered related to the high LUMO level of the fluorescent dopant comparatively to that of the phosphorescent dopant. Regarding the device efficiencies, best performances were achieved while separating the two doped layers by a thin bilayer of the two hosts stacked on top of each other. A maximum CE and PE of 14.1 cd/A and 7.5 lm/W, a brightness of 2700 cd/m^2 at 20 mA/cm^2 were respectively obtained with EB-502 as the blue fluorescent dopant and $Ir(piq)_2(acac)$ as the phosphorescent dopant. Energy transfer from the fluorescent to the phosphorescent dopant was efficiently prevented, avoiding singlet energy transfer from the wide bandgap dopant to the low bandgap dopant. Efficiency roll-off was merely 6% between 5 and 30 mA/cm^2. However, considering that the LUMO of the fluorescent host was too close to that of the HTL, enabling the electron to pass in the HTL, 9,9-*bis*-[4-(*N,N-bis*-biphenyl-4-ylamino)phenyl]-9H-fluorene (BPAPF) was substituted by TAPC. The efficient confinement of electrons on the dopant increased the current and power efficiencies till 16.3 cd/A and 8.1 lm/W and broaden the RZ, evidencing the electron leakage in the case of BPAPF. Likewise, a hybrid structure was proposed to regulate the distribution of charge carriers in EMLs and thus stabilizing the spectrum emission of the fabricated hybrid WOLEDs [98]. To efficiently balance the hole and electron transportation within the EML, an appropriate mixture of TCTA and $Bepp_2$ was used to host the red and the green phosphorescent emitters into two adjacent layers. Using this bipolar host, the width of exciton recombination region could be extended to the whole red and green-emitting layers. The effective control of charge carriers in EMLs was achieved by introducing an interlayer between the green and the blue fluorescent dopant, controlling the electron injection in the red and green EMLs and reducing the space charge accumulation. WOLEDs with excellent spectral stability could be obtained and the CE and PE reached 30.8 cd/A and 27.1 lm/W at the luminance of 1000 cd/m^2.

5. Devices with Phosphorescence-Interlayer-Phosphorescence (PIP EML) Emissive Structure

A logical extension of the previous works (FIP EML) consisted in separating two layers doped with a phosphorescent emitter by an interlayer. Only few investigations were devoted to the design of ultrahin layer-based devices [99, 100]. One of the major breakthrough in this field was published in 2008 by introducing an ultrathin interlayer of an orange phosphorescent dopant between the double emissive layer of blue PhOLEDs [101]. Especially, insertion of a 0.25 nm thick layer of 3 wt% iridium(III) *bis*-(2-phenylquinoly-$N,C^{2'}$)dipivaloylmethane PQ_2Ir-doped TCTA and a 0.25 nm thick layer of 3 wt% PQ_2Ir-doped 26DCzPPy between the two blue emissive layers were sufficient to induce an orange emission and achieve the desired white emission. Only minor changes in the color with the driving voltage were observed, denoting that no significant drift of the recombination zone occured. However, if a relative stability of the emission color was evidenced, the PE of 53 lm/W at the luminance of 100 cd/m² rolled off to 44 lm/W at 1 000 cd/m², corresponding to a 17% decrease. Recently, this concept was extended to devices comprising iridium and platinum complexes as co-emitters [102]. Using the following device configuration ITO/ 4,4,4-{*N*-(2-naphthyl)-*N*-phenylamino}-triphenylamine (2T-NATA) (50 nm)/ NPB (5 nm)/ TCTA (8 nm)/ mCP:2 wt% platinum [1,3-difluoro-4,6-di(2-pyridinyl)benzene]chloride (Pt-4) (8 nm)/ TPBI: 5 wt% PQIr (0.25 nm)/ TPBI (0.5 nm)/ TPBI: 6 wt% $Ir(ppy)_3$ (0.75 nm)/ TPBI (30 nm)/ LiF (0.3 nm)/ Al), the operating voltage could be markedly reduced, compared to the reference device with thicker layers, as a result of the ultrathin layers. A PE of 11 lm/W was thus reached at 1000 cd/m² contrarily to the 3.6 lm/W only obtained with the reference devices.

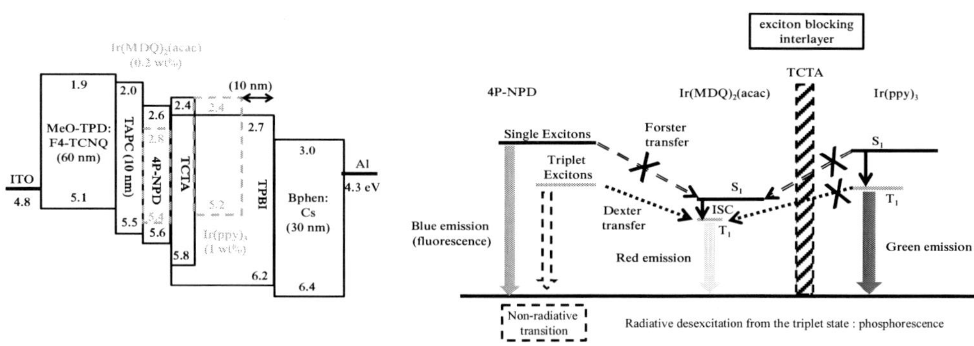

Figure 14. Structure allowing to recycle the non-emissive triplet excitons of the fluorescent emitter to radiative excitons by Dexter energy transfer to a phosphorescent one.

Concomitantly, other authors reported the introduction of an exciton balancing interlayer between two EMLs comprising a phosphorescent dopant that greatly improved the chromaticity [103]. By harvesting the non-emissive triplet excitons of the fluorescent blue emitter, a reduced efficiency roll-off was obtained. Here again, recycling of the non-emissive triplet excitons is based on the specific diffusion length and the exceptional lifetime of these excitons.

The key-step consisted to directly dope the orange phosphor in the fluorescent blue layer, such that the extremely fast energy transfer of the triplet excitons from the fluorescent emitter to the phosphor could severely reduce the concentration quenching (see Figure 14). However, to maintain the blue emission of the fluorophor and thus produce a white light, a certain minimal thickness of this layer was required. To avoid a possible Förster energy transfer of the 4P-NPD singlets to $Ir(MDQ)_2(acac)$, a extremely low doping concentration was used. High power and current efficiencies reported for this simple two-wavelength phosphorescent/fluorescent WOLED can also be assigned to the use of a light extraction enhancing microlens foil. Light outcoupling enabled to increase the PE from the initial value of 25.4 lm/W to 32.3 lm/W at 1000 cd/m², corresponding to a net 27% increase.

6. Suppressing the Efficiency Roll-Off by Triplet to Singlet Exciton Conversion

Highly efficient OLEDs with low roll-off can only be obtained if the electrical loss by charge imbalance, exciton-polaron quenching and triplet-triplet annihilation becomes negligible. Triplet-triplet annihilation is well-established as originating from the long lifetime of triplet excited state of the dopant and the long-living triplet excited state lifetime of heavy transition metal complexes has been for long the main key adverse factor to incorporate these complexes in OLEDs.

An innovative strategy has recently been developed, based on a co-doped emissive layer comprising a phosphor ($Ir(ppy)_3$) and a fluorophor (*N,N'*-dimethylquinacridone (DMQA)) (see Figure 15) [104]. In this co-doped strategy, the triplet excitons formed in the phosphor are efficiently converted as singlet excitons on the fluorophore by energy transfer, thus significantly reducing the $Ir(ppy)_3$ exciton lifetime and suppressing the triplet–triplet annihilation and triplet-polaron annihilation. If the concept of energy transfer from the triplet state of the dopant to the singlet state of the co-dopant was clearly demonstrated, device performances remained however limited, with a maximal EQE only reaching 9.3% after optimization.

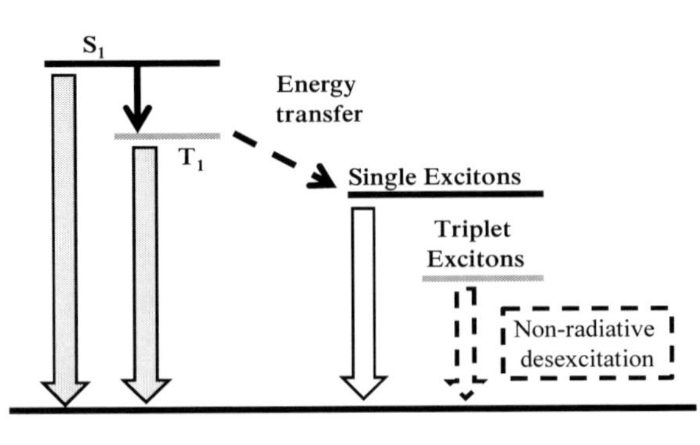

Figure 15. Excitons management in OLEDs co-doped with a fluorescent and a phosphorescent dopant.

7. Solving the Charge Imbalance by the Convenient Functionalization of the Dopants

In all the above-mentioned articles, reduction of the efficiency roll-off has been obtained by improving the charge balance of the host, by mean of blending a hole and an electron transporters, or by forming a double emissive layer, one layer comprising the hole transporter as host, the other one, an electron transporter as host. A good confinement of excitons was obtained. Chemical engineering around the development of bipolar hosts has also been carried out to improve charge balance. In addition to these strategies, an alternative approach consisted in improving the electron transport ability of the dopant by the convenient substitution with electron carriers [105]. It is well-established that the HOMO level of heteroleptic iridium complexes $Ir(C^{\wedge}N)_2(LX)$ is essentially metal-centered and that the LUMO level is mostly localized on the cyclometalated ligand. In this regards, the most widely used cyclometallated ligand i.e., 2-phenylpyridine has been extensively modified and a wide range of derivatives have been proposed [106]. The ancillary ligand is not involved in both the HOMO and LUMO level but it can significantly alter the excited state lifetime [107]. Ability of tetraphenylimidodiphosphinate acid (Htpip) to improve the electron-transporting ability of the dopant was clearly evidenced. Two complexes were synthesized, namely $Ir(tfmppy)_2(tpip)$ and $Ir(dfppy)_2(tpip)$. While doping mCP at 6 wt% with $Ir(dfppy)_2(tpip)$, a maximum CE and PE of 67.95 cd/A and 69.90 lm/W were respectively obtained. A brightness of 64 351 cd/m^2 at 11.8 V was also observed. The roll-off ratio between 100 and 1 000 cd/m^2 was 11.92%, close to that reported by other groups. The low LUMO level of the two complexes was determined as being beneficial and favoring the electron transport, even at high dopant concentration. By the balanced hole-electron transport ability of the EML, recombination probability of electrons and holes was greatly improved, with broadening of the recombination zone. Still for diminishing the intermolecular interactions of the phosphorescent molecules and alleviating the undesired self-quenching, the construction of a series of iridium complexes bearing benzothiazoles as sterically hindered ligands was studied [108]. *Bis*[2-(3′,5′-di-*tert*-butylbiphenyl-4-yl)benzothiazolato-*N*,$C^{2'}$] iridium(III) acetylacetonate $(tbpbt)_2Ir(acac)$ and *bis*[2-(9,9-dimethyl-9*H*-fluoren-2-yl)benzothiazolato-*N*,$C^{2'}$] iridium (III) acetylacetonate $(fbt)_2Ir(acac)$ were both evaluated as phosphors with suppressed molecular interactions. The two complexes were studied in a conventional configuration i.e., ITO/ NPB (50 nm)/ Ir complex:CBP (30 nm)/ BCP (10 nm)/ Alq_3 (40 nm)/ Mg:Ag (10:1, 100 nm) with a dopant ratio varying from 7 to 100%. Best performances were achieved at 15 wt% doping level for the two dopants. A maximum brightness, CE and EQE of ~30,000 cd/m^2, 26.9 cd/A, and 9.44%, and ~26,000 cd/m^2, 28.5 cd/A, and 15.6% were respectively obtained for $(tbpbt)_2Ir(acac)$ and $(fbt)_2Ir(acac)$. Although efficiency decreased with increasing the current density, no severe efficiency roll-off was observed. If the steric hindrance of the two complexes was almost similar, lower performances of $(tbpbt)_2Ir(acac)$-based devices were partly ascribed to the presence of the *tert*-butyl substituents on $(tbpbt)_2Ir(acac)$ that may lead to rather inferior carrier mobility due to the electrical insulating character of these substituents. As it could be anticipated, substitution of iridium complexes by carbazoles was studied [109]. Previous works devoted to the substitution of emitters with hole-transporting groups seemed to be a wise choice as the improvements of device performances were really encouraging [110]. In the present case, enhancement of the device performances with these modified emitters can't be contested with regards of the results obtained with the parent complexes free of this substitution. Performance of the substituted *bis*[2-(4-(4-(9*H*-carbazol-9-yl)butoxy)phenyl)benzo[*d*]thiazole-*N,C*$^{2'}$] iridium(III) acetylacetonate $(cbbt)_2Ir(acac)$ were much higher than that obtained with the parent $Ir(bt)_2(acac)$ complex as exemplified with the maximal EQE of 10.3% and a maximal CE of 25.5 cd/A obtained after substitution, compared to the values of 9.7% and 9.6 cd/A obtained without substitution.

Recently, concentration insensitive OLEDs were designed with a bipolar phosphorescent emitter [111]. This specific feature is of crucial interest as the precise control of the dopant concentration in the EML (and the adjacent layers) determines the device performances. However, this extreme precision can be challenging to achieve, especially for large scale manufacturing process. Insensitivity could greatly help the device elaboration. To date, only few reports have been devoted to concentration-insensitive devices, with inequal successes concerning the roll-off issue [112]. In the present exemple, the simplification also concerned the device architecture as PhOLEDs were reduced to trilayer devices. Using iridium (III) *bis*(2-phenylpyridinato)-*N,N'*-diisopropyl-diisopropyl-guanidinate $(ppy)_2Ir(dipig)$ as the emitter, the EQE roll-off values were lower than 3% at 1 000 cd/m² and 15% at 10 000 cd/m². Very high maximum PE and EQE were obtained since values as high as 98.5 lm/W and 26.5% at 15 wt% dopant concentration and 98.1 lm/W and 25.4% at 30 wt% were respectively observed. These high and stable efficiencies were assigned to the bipolar character of the dopant, favoring the charge recombination in the whole layer. Still with aim of simplying the device elaboration, the convenient functionalization of emitters can avoid the use of hosts while offering the opportunity of fabricating solution-processable devices [113]. The balanced charge transportation were furnished by the electron-transport 1,3,4-oxadiazole and the hole-transport ability was provided by the carbazole moiety. Devices performances increased with the number of oxadiazole and carbazole moieties per complex as well as the respect of the parity between the holes and electrons transport moieties. The tetrasubstituted complex $IrPBIC_2O_2$ was determined as being the best complex in subduing the quenching effect in the solid state by its inherent steric hindrance. A maximum brightness of 6255 cd/m² at 4.9 V, maximum CE and PE respectively of 15.41 cd/A and 8.08 lm/W, an EQE of 4.68% were obtained with $IrPBIC_2O_2$.

8. UNHOSTED PHOLEDS

Other approaches such as modifying the structure of the emissive layer have also been used in efforts to overcome triplet-triplet annihilation and reducing the efficiency roll-off [114]. Without taking recourse to complex strategies such as blending and doping, only few examples of unhosted devices are reported in the literature. However, devices based on ultrathin emissive layers instead of doped structures are easier to fabricate by avoiding the difficult step of co-evaporation thus favoring the reproducibility of the device fabrication. Phosphorescent emitters were seldomly reported as ultrathin layer materials, contrarily to the fluorescent materials [115]. As first example, *bis*(2-(4-*tert*-butylphenyl)benzothiazolato-*N,C*$^{2'}$) iridium(III) acetylacetonate $(t\text{-}bt)_2Ir(acac)$ that exhibit bulky *tert*-butylphenyl substituents and a low ionization potential was studied to fabricated nondoped devices. Two configurations were investigated i.e., ITO/ NPB (50 nm)/ $(t\text{-}bt)_2Ir(acac)$ (1 nm)/ NPB (x nm)/ BCP (30 nm)/ Mg:Ag (x=0 for device A and x=6 for device B), differing by the presence or not of a NPB interlayer (see Figure 16). Lower performances of devices A only comprising the thin phosphorescent emissive layer was assigned to the efficient charge trapping done by the complex, originating from the strong energy gaps of its respective HOMO and LUMO levels with those of the adjacent layers. On the opposite, in devices B, the NPB interlayer efficiently blocked electrons by its hole only transport ability, and modified the EL process. In this case, holes drift until the NPB interlayer in which non-emitting triplet excitons formed. Due to the long diffusion length of triplet excitons, these non-emitting excitons can diffuse into the ultrathin emissive layer. In addition of this first EL process, electrons can also cross the NPB layer and be directly injected in the LUMO level of the dopant, thus constituting a second EL pathway.

This initial work, done on a yellow emitter, was further extended to a green one, where CBP was used this time as HTL and as interlayer [116]. The device configuration was the following : ITO/ NPB (30 nm)/ TAPC (20 nm)/ CBP (15 nm)/ 1,2-dipheny1-1*H*-benzo-imidazole)iridium (acetylacetonate) $(pbi)_2Ir(acac)$ (0.2 nm)/ CBP (15nm)/ BPhen (40 nm)/ Mg:Ag. As specificity, excimer emission dominated the EL spectra, red-shifting the emission of 16 nm compared to the PL spectrum of the complex. Devices performances steadily decreased while increasing the EML thickness, evidencing quenching effects such as triplet-triplet annihilation (TTA) and triplet-polaron annihilation (TPA) to become dominant. By decreasing the EBL (TAPC) thickness, formation of triplet excimers could be efficiently suppressed, less electrons being confined within the EML. At this optimum thickness of 15 nm for EBL, a maximum luminance of 26 531 cd/m^2, a CE of 36.2 cd/A, and a PE of 32.4 lm/W were obtained. Effective reduction of the luminescence quenching caused by the aggregation of the emissive species can be achieved if each complex can be efficiently separated from each other. To maximize the isolation of the emissive centers in host-free devices, dendronized iridium (III) complexes were investigated as a possible alternative [117].

If the effectiveness of the separation was demonstrated, this strategy suffered as main drawback from an extensive work of synthesis that has to be done prior to device fabrication. Complexation of the dendronized ligands to iridium and preparation of the corresponding homoleptic complexes is often discouraging considering the low reaction yields ranging from 10 to 35% yield, as a result of the steric hindrance and the low solubility of the bulky dendritic ligands. To investigate the effectiveness of this self-host strategy, enlargement of the dendritic arms from the first (G1) to the third generation (G3) was studied. The relationship between dendrimer's generation and device performance was examined in the following structure: ITO/ PEDOT:PSS (50 nm)/ Ir-G1, Ir-G2, or Ir-G3 (80 nm)/ TPBI (25 nm)/ Ba (4 nm)/ Al. All devices exhibited a low turn-on voltage (2.8-3.9 V) assigned to the low HOMO levels of the three dendrimers close to the workfunction of the hole injection layer. Best performances were obtained with Ir-G1. A maximum CE of 40.9 cd/A, a maximum PE of 39.5 lm/W, and a maximum EQE of 16.4% were obtained. While examining the performances of dendrimers of superior generation, a severe drop of efficiency was observed, diminishing to 7.6 cd/A, 5.4 lm/W and 3.0% for the current, power efficiencies as well as the EQE respectively for Ir-G3. This unexpected behavior was assigned to triplet excitons annihilation on TPA units, the number of external TPA units sharply increasing from 30 units for Ir-G2 to nearly 300 units for Ir-G3. Low efficiency roll-off for the best devices (Ir-G1) was assigned to the sufficiently depressed concentration quenching of the emitters and the good charge-transporting ability of the dendrons.

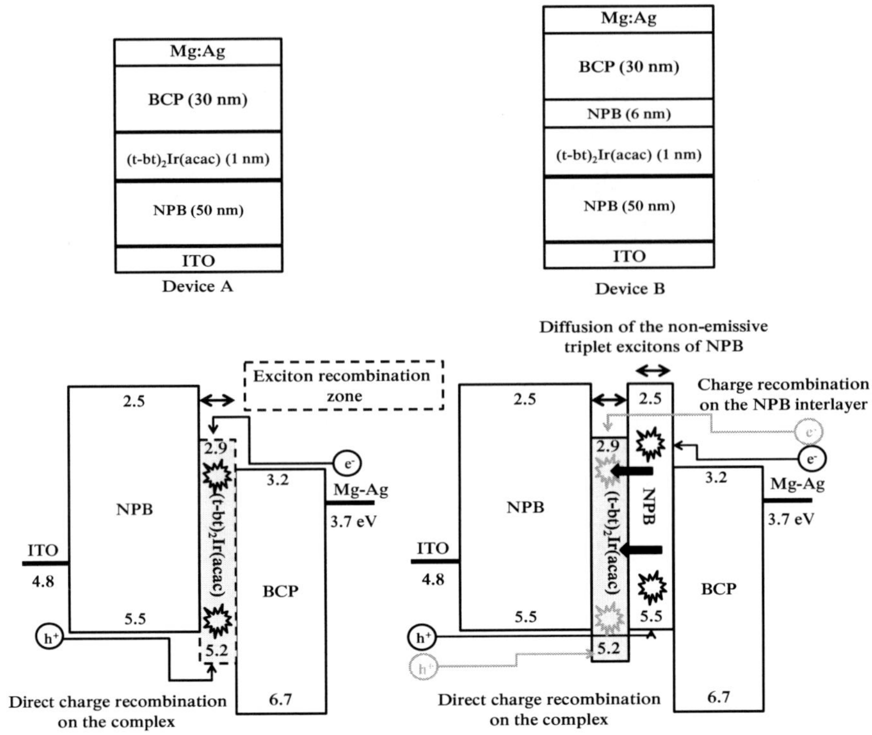

Figure 16. OLEDs comprising ultra-thin layers of unhosted phosphorescent emitters.

9. Use of a Coordination Complex As Host for Phosphorescent Dopants

To date, metal complexes have only been scarcely tested as hosts for OLEDs. The first report mentioning the use of a coordination complex as host for a phosphorescent dopant was reported last year [118]. In this aim, two complexes were studied as potential hosts, namely $Be(PPI)_2$ and $Zn(PPI)_2$ (where PPI stands for 2-(1-phenyl-1*H*-phenanthro [9,10-*d*]imidazol-2-yl)phenol) that both exhibits the same HOMO and LUMO levels. Interest of this unprecedented strategy is that the host benefits from the tridimensional and rigid molecular structure inherent to coordination complexes, as well as from a higher thermal stability, as commonly observed for metal complexes. The host also inherits from the intrinsic carrier-transport ability of metal complexes. In term of thermal stability, *Tg* of $Zn(PPI)_2$ was observed at 217°C, largely higher than that observed for organic small-molecules. Two iridium (III) complexes were tested as dopants, i.e., a green dopant $Ir(ppy)_3$ and a red dopant $Ir(MDQ)_2(acac)$. $Be(PPI)_2$-based devices proved to exhibit higher performances than $Zn(PPI)_2$-based devices with $Ir(ppy)_3$ as dopant, especially for addressing the roll-off issue. EQE efficiency of $Be(PPI)_2$-based devices decreases slightly with increasing the current density with only ca. 3.3% roll-off at 1000 cd/m² relative to the maximum EQE, whereas a decrease of 12.5% was observed for $Zn(PPI)_2$-based devices. Conversely, similar performances were obtained with the red dopant, clearly evidencing the intrinsic bipolar nature of the two host complexes. Given this, wide bandgap of the two complexes was also favorable to a facile charge injection from the adjacent layer, thereby improving the device performances. Benefiting from all these factors, $Be(PPI)_2$-based devices exhibited a maximum luminance of 92 261 cd/m², a CE and a PE of 55.6 cd/m² and 61.4 lm/W respectively with $Ir(ppy)_3$ as the dopant. Recently, one of the most common green phosphorescent dopant, namely $Ir(ppy)_3$, was used for the first time as a co-host for a phosphorescent yellow dopant PO-01 [119]. Comparison with devices only comprising the organic host showed a slight efficiency increment of 1%, from 42 to 47 lm/W, assigned to the trapping effect of the organometallic host, resulting in a decrease of the current density. Improvement of the device performance was also attributed to the parallel desexcitation pathways issued from the presence of the organometallic co-host. First, singlet and triplet excitons formed on the organic host can be efficiently transfered to the yellow dopant, as in conventional OLEDs. As a second desexcitation pathway, part of these excitons generated on the organic host can also be transfered to the organometallic co-host. In this last case, the desexcitation mechanism is accelerated, by the shorter excited state lifetime of the organometallic complexes relatives to those of the organic molecules. In turn, the resultant yellow OLEDs showed a PE of 59 lm/W, a CE of 71 cd/A at 1000 cd/m². Parallel desexcitation pathways also contributed to significantly reduce the exciton concentration. Devices comprising a combination of an organic and an organometallic host ($Ir(ppy)_3$) for the sensitization of a fluorescent red dopant were previously reported for the design of hybrid WOLEDs, in which the co-doped green/red layer was inserted between two blue-emitting layers, the different emissive layers being separated by a thin organic host interlayer (see Figure 17) [120].

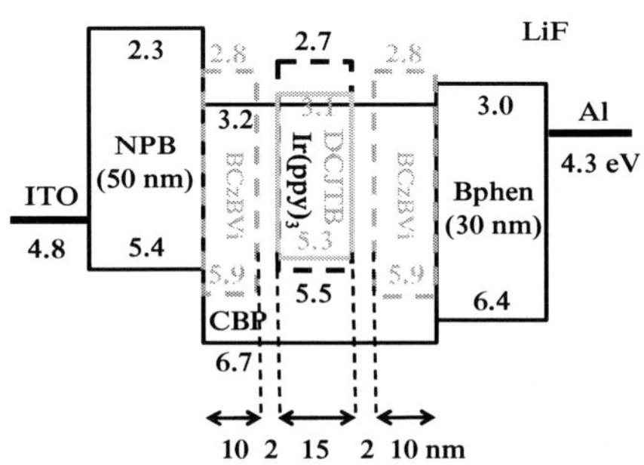

Figure 17. OLED with two thin organic host interlayers used to control the diffusionrate of the non-emissive triplet excitons of the blue dopant towards the red emitter.

By lightly doping with a red emitter, uncomplete energy transfer from the organometallic host was observed, resulting in a mixture of red and green emissions. The red emission from DCJTB was also determined as increasing with elevating the $Ir(ppy)_3$ concentration. By increasing the $Ir(ppy)_3$ concentration and by phosphorescent sensitization, the non-emissive triplet excitons formed on the blue fluorescent dye can diffuse through the CBP interlayer and efficiently harvested by the red dopant for light emission.

10. Exciton Management by Harvesting the Triplet Excitons of Fluorescent Emitters

A key requirement for achieving high emission in OLEDs is that all triplet excitons are utilized for light emission. This is a major limitation, especially for fluorescent materials, as only the singlet excitons can be harvested for light emission. However, for devices combining a fluorescent and a phosphorescent emitter, it is possible for the phosphorescent dopant to recycle the triplet excitons of the fluorescent emitter for light generation. The first results in this field were obtained in 2006 by Thompson and co-workers [121]. In this work, triplet and singlet excitons of the host (CBP) were both harvested along independent channels, the singlet excitons being harvested by the fluorescent dopants and the triplet excitons by the phosphorescent dopants, all Förster energy transfer being impeded by the presence of a thin interlayer of CBP.

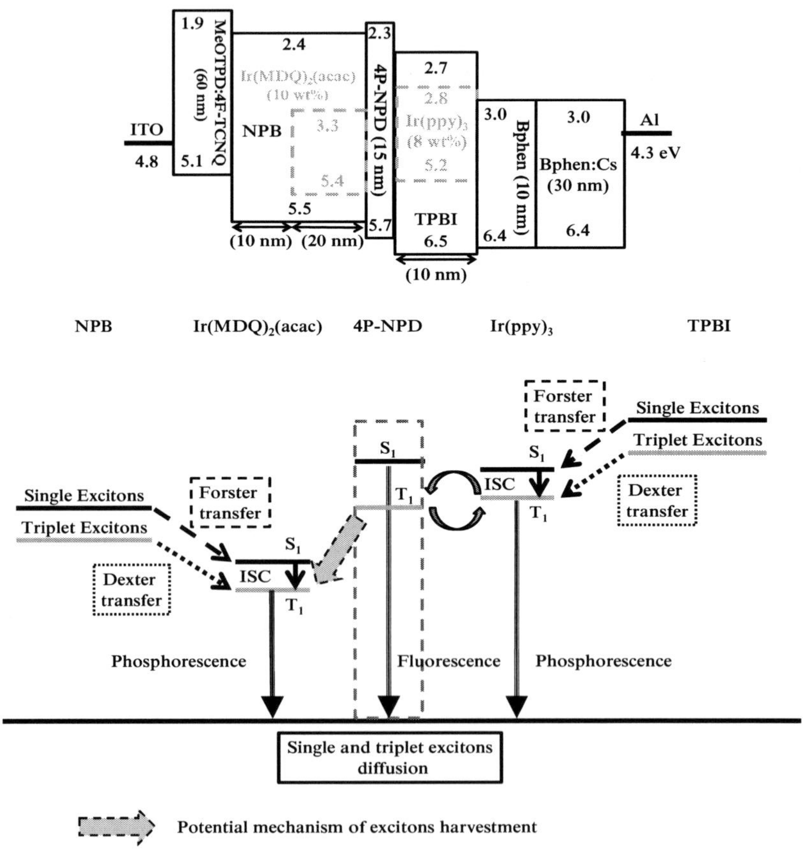

Figure 18. Devices harvesting the non-emissive triplet excitons of the blue fluorescent emitter 4P-NPD.

This initial and exciting report was followed by other investigations in which the level of sophistication was increased until utilizing the concepts of exciton management for efficient white light generation (See Figure 18) [122]. In this concept, a blue fluorescent emitter (4P-NPD) is introduced as an interlayer between one EML comprising the phosphorescent green dopant $Ir(ppy)_3$ and another one comprising the phosphorescent orange-red emitter $Ir(MDQ)_2(acac)$. Efficiency of the cascade energy transfer is directly related to the energy gaps of the three emitters and their correct alignments in each layer. First, all excitons formed on the blue fluorescent emitter (4P-NPD) are collected, the singlet ones decaying radiatively for blue emission, the non-emissive triplet excitons of 4P-NPD being harvested (Dexter transfer) by the phosphorescent green dopant in the adjacent layer. As specifity, triplet energy level of $Ir(ppy)_3$ is close to that of 4P-NPD. Transfer of the triplet excitons from 4P-NPD to the green phosphor is rendered possible by the long diffusion length and the long lifetime of these excitons, compared to the singlet excitons. Eventually, part of the triplet excitons formed on 4P-NPD can also diffuse towards the NPD layer where they can still be harvested by the orange-red emitter $Ir(MDQ)_2(acac)$.

11. Exciton Management and Phosphorescent Sensitization

There are a number of ways to realize WOLEDs. In the previous approach, the non-emissive triplet excitons of the fluorophor were recycled by the phosphorescent emitter to produce light. One possible extension of this strategy consists in also utilizing part of the triplet excitons of the phosphorescent emitter for sensitizing another emitter. This new route combining exciton management and phosphorescent sensitization enabled to design WOLEDs achieving almost 100% internal quantum efficiency with a severe reduction of the efficiency roll-off (see Figure 19) [123]. More specifically, $Ir(ppy)_3$ was sensitized by a blue fluorophor (NPB) introduced in an adjacent layer, this phosphor being itself used as an emitter and as a sensitizer for an orange fluorophor (DCJTB) introduced in the same layer. In this device ITO/ PEDOT:PSS (30 nm)/ PVK:$Ir(ppy)_3$:DCJTB (100:5:0.4 in wt.) (60 nm)/ NPB (4 nm)/ BCP (10 nm)/ Alq_3 (20 nm)/ LiF (1 nm)/Al, thickness of the NPB layer was determined as being crucial and the optimum PE was obtained for a thickness of 4 nm. Although these WOLEDs gave less pronounced efficiency roll-off, the obtained maximum brightness remained limited to 8 700 cd/m^2 at 210 mA/cm^2, corresponding to a PE of 4 lm/W and an EQE of 6.9%. A deteriorated effect on the color stability of WOLEDs was also evidenced for thickness higher than 4 nm, resulting from a shift of the exciton recombination zone with increasing the operating voltage. As compared to the all-fluorescence WOLEDs reported before, these devices combining a dual phosphorescent/fluorescent sensitization exhibited performances similar to that obtained with traditional fluorescence-based WOLEDs [124]. Parallel to this work, other conceptual advancements led to several types of phosphorescent/fluorescent architectures, leading to higher device performances. Notably, combination of a fluorescent blue emitter (DSA-Ph) and an orange phosphor *bis*(2-(9-phenylcarbazolyl)-2-pyridyl) iridium(III) acetylacetonate $Ir(Cz\text{-}CF_3)$ enabled to fabricate WOLEDs yielding a current efficiency of 42.5 cd/A and a power efficiency of 24.9 lm/W [110g]. To reach these outstanding performances, the blue and orange dopants were introduced into two adjacent layers, separated by a thin HTL interlayer. Using this strategy, electrons were efficiently blocked in the fluorescent layer by the presence of the interlayer. Jointly, the thin hole-transporting interlayer facilitated the hole injection, favoring the charge recombination in this layer and guaranteing the blue emission of the fluorophor. Thanks to their long diffusion lengths, the non-radiative triplets excitons generated on the fluorophor could migrate from their formation zone to the phosphor-doped layer, ensuring the orange emission. Singlets and triplets were thus efficiently channeled to their respective emitting species. Among the paramaters influencing the color stability, increase of the thickness of the phosphorescent layer resulted in an orange emission overhelming the blue emission. Device performances were greatly enhanced by the presence of a PEDOT:PSS buffer layer, lowering the hole-injection barrier and reducing the driving voltage.In this example, the orange emission was obtained with a unique emitter. The same strategy can also be developped with a mixture of a red and a green emitter to furnish the orange emission [125]. Hosts for the fluorescent and the phosphorescent dopants can also be replaced by a mixture of co-hosts. Given these considerations, hybrid structures using a blue fluorescent emitter and red/green phosphorescent dopants codoped in a mixed host proved to be an effective method to reduce the efficiency roll-off at high luminance and enhance the device color-stability. Color coordinates of WOLEDs were notably kept constant up to luminance of 10 000 cd/m², with a decrease of the efficiency only corresponding to 2% of the maximum quantum efficiency.

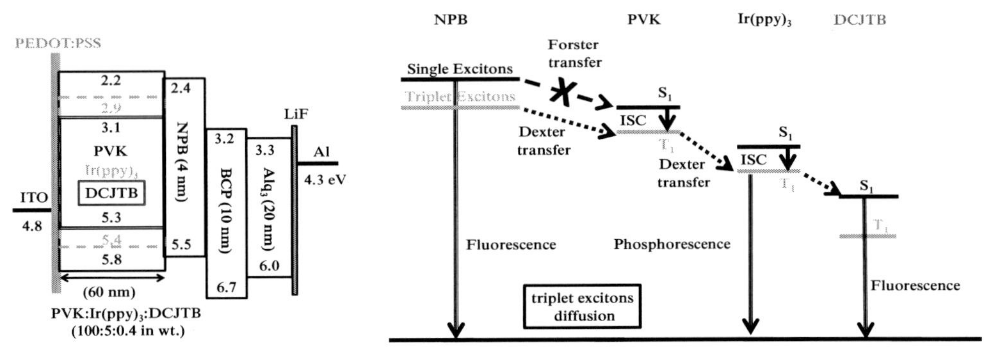

Figure 19. OLEDs based on a cascade energy transfer of triplet excitons.

12. Electrophosphorescent Inverted Top-Inverted Top-Emitting Organic Light-Emitting Devices

As final and effective strategy, electrophosphorescent inverted top-emitting organic light-emitting diodes can also address the roll-off issue [126]. By modifying the workfunction of the top electrode with HAT-CN (1,4,5,8,9,11-hexaazatriphenylene hexacarbonitrile), green PhOLEDs with a CE of 119 cd/A were obtained at 100 cd/m². As appealing features, the CE only rolls off to 110 cd/A at 1000 cd/m² and devices exhibited a turn-on voltage of only 3.0 V. It is worth noting that an optical outcoupling layer was deposited on top of the Ag anode, explaining the unusual high current efficiency.

Conclusion

Mixed hosts and multilayer device structures are suggested as the most promising structures to maximize the current efficiency and to address the quantum efficiency roll-off issue. Using this strategy, blue, green or white PhOLEDs with low roll-off characteristic up to 10 000 cd/m² were successfully developed. However, such sophisticated device architectures inevitably increase the manufacturing complexity and the production cost. The concern of the reproducibility of device performances is another adverse factor to the development of these complicated devices. Parallel to mixed hosts, devices with double EML, each layer comprising its own host matrice, have also been explored as a possible alternative to the blend approach. Finally, the simplest strategy to suppress the efficiency roll-off at high current density definitely consists in the development of new hosts with bipolar character that could equally inject and transport holes and electrons within the EML. However, in this aim, an extensive work to synthesize these new materials has to be carried out by chemists beforehand. After addressing the roll-off issue, the simplification of device structure should become the next challenge for the future. After overcoming a decade of obstacles, organic devices could become mature to reach commercialization.

References

[1] a) Service, R.D. Organic LEDs look forward to a bright, white future. *Science* 2005, 310, 1762–1763 b) Reineke, S.; Lindner, F.; Schwartz, G.; Seidler, N.; Walzer, K.; Lüssem, B.; Leo, K. White organic light-emitting diodes with fluorescent tube efficiency. *Nature* 2009, 459, 234–238 c) Tang, C.W.; Van Slyke, S.A. Organic electroluminescent diodes. *Appl. Phys. Lett.* 1987, 51, 913–915 d) Burroughes, J.H.; Bradley, D.D.C.; Brown, A.R.; Marks, R.N.; Mackay, K.; Friend, R.H.; Burns, P.L.; Holmes, A.B. Light-emitting diodes based on conjugated polymers. *Nature* 1990, 347, 539–541 e) Forrest, S.R. The path to ubiquitous and low-cost organic electronic appliances on plastic. *Nature* 2004, 428, 911–918 f) So, F.; Kido, J.; Burrows, P. Organic light-emitting devices for solid-state lighting. *MRS Bull.* 2008, 33, 663–669 g) Kelley, T.W.; Baude, P.F.; Gerlach, C.; Ender, D.E.; Muyres, D.; Haase, M.A.; Vogel, D.E.; Theiss, S.D. Recent progress in organic electronics: materials, devices, and processes. *Chem. Mater.* 2004, 16, 4413–4422.

[2] a) Adachi, C.; Baldo, M.A.; Thompson, M.E.; Forrest, S.R. Nearly 100% internal phosphorescence efficiency in an organic light emitting device. *J. Appl. Phys.* 2001, 90, 5048–5051 b) Tung, Y.-L.; Lee, S.-W.; Chi, Y.; Tao, Y.-T.; Chien, C.-H.; Cheng, Y.-M.; Chou, P.-T.; Peng, S.-M.; Liu, C.-S. Organic light-emitting diodes based on charge-neutral Os (II) emitters: generation of saturated red emission with very high external quantum efficiency. *J. Mater. Chem.* 2005, 15, 460–464 c) Xiao, L.; Su, S.-J.; Agata, Y.; Lan, H.; Kido, J. Nearly 100% internal quantum efficiency in an organic blue-light electrophosphorescent device using a weak electron transporting material with a wide energy gap. *Adv. Mater.* 2009, 21, 1271–1274 d) Lamansky, S.; Djurovich, P.; Murphy, D.; Abdel-Razzaq, F.; Lee, H.-E.; Adachi, C.; Burrows, P.E.; Forrest, S.R.; Thompson, M.E. Highly phosphorescent *bis*-cyclometalated iridium complexes: synthesis,

photophysical characterization, and use in organic light emitting diodes. *J. Am. Chem. Soc.* 2001, 123, 4304–4312.

[3] Ho, C.-L.; Wong, W.-Y. Small-molecular blue phosphorescent dyes for organic light-emitting devices. *New J. Chem.* 2013, 37, 1665–1683.

[4] a) Baldo, M.A.; O'Brien, D.F.; You, Y.; Shoustikov, A.; Sibley, S.; Thompson, M.E.; Forrest, S.R. Highly efficient phosphorescent emission from organic electroluminescent devices. *Nature* 1998, 395, 151–154 b) Wong, W.-Y.; He, Z.; So, S.-K.; Tong, K.-L.; Lin, Z. A multifunctional platinum-based triplet emitter for OLED applications. *Organometallics* 2005 24, 4079–4082.

[5] a) Tung, Y.-L.; Wu, P.-C.; Liu, C.-S.; Chi, Y.; Yu, J.-K.; Hu, Y.-H.; Chou, P.-T.; Peng, S.-M.; Lee, G.-H.; Tao, Y.; Carty, A.J.; Shu, C.-F.; Wu, F.-I. Highly efficient red phosphorescent osmium (II) complexes for OLED applications. *Organometallics* 2004, 23, 3745–3748 b) Carlson, B.; Phelan, G.D.; Kaminsky, W.; Dalton, L.; Jiang, X.; Liu, S.; Jen, A. K.-Y. Divalent osmium complexes: synthesis, characterization, strong red phosphorescence, and electrophosphorescence. *J. Am. Chem. Soc.* 2002, 124, 14162–14172 c) Lee, T.-C.; Hung, J.-Y.; Chi, Y.; Cheng, Y.-M.; Lee, G.-H.; Chou, P.-T.; Chen, C.-C.; Chang, C.-H.; Wu, C.-C. Rational design of charge-neutral, near-infrared-emitting osmium (II) complexes and OLED fabrication. *Adv. Funct. Mater.* 2009, 19, 2639–2647 d) Chou, P.-T.; Chi, Y. Osmium- and ruthenium-based phosphorescent materials: design, photophysics, and utilization in OLED fabrication. *Eur. J. Inorg. Chem.* 2006, 17, 3319–3332.

[6] a) Bernards, D.A.; Biegala, T.; Samuels, Z.A.; Slinker, J.D.; Malliaras, G.G.; Flores-Torres, S.; Abruna, H.D.; Rogers, J.A. Organic light-emitting devices with laminated top contacts. *Appl. Phys. Lett.* 2004, 84, 3675–3677 b) Shahroosvand, H.; Najafi, L.; Mohajerani, E.; Janghouri, M.; Nasrollahzadeh, M. Unusual electroluminescence in ruthenium(II) tetrazole complexes. *RSC Adv.* 2013, 3, 6323–6326.

[7] a) Ulbricht, C.; Beye, B.; Friebe, C.; Winter, A.; Schubert, U.S. Recent developments in the application of phosphorescent iridium (III) complex systems. *Adv. Mater.* 2009, 21, 4418–4441 b) Diouf, B.; Jeon, W.S.; Pode, R.; Kwon, J.H. Efficiency control in iridium complex-based phosphorescent light-emitting diodes. *Adv. Mater. Sci Eng* 2012, 794674-1–794674-14 c) Dumur, F.; Lepeltier, M.; Zamani Siboni, H.; Gigmes, D.; Aziz, H. Highly efficient organic light-emitting devices prepared with a phosphorescent heteroleptic iridium (III) complex containing 7,8-benzoquinoline as the cyclometalated ligand, *Adv. Opt. Mater.* 2015, 2, 262-266.

[8] a) Kohler, A.; Wilson, J.S.; Friend, R.H. Fluorescence and phosphorescence in organic materials. *Adv. Mater.* 2002, 14, 701–707 b) Dedeian, K.; Shi, J.; Shepherd, N.; Forsythe, E.; Morton, D.C. Photophysical and electrochemical properties of heteroleptic *tris*-cyclometalated iridium (III) complexes. *Inorg Chem.* 2005, 44, 4445–4447 c) Luo, Y.; Aziz H. Correlation between triplet–triplet annihilation and electroluminescence efficiency in doped fluorescent organic light-emitting devices. *Adv. Funct. Mater.* 2010, 20, 1285–1293.

[9] Wu, C.; Chen, H.-F.; Wong, K.-T.; Thompson, M.E. Study of ion-paired iridium complexes (soft salts) and their application in organic light emitting diodes. *J. Am. Chem. Soc.* 2010, 132, 3133–3139.

[10] a) Dumur, F.; Nasr, G.; Wantz, G.; Mayer, C.R.; Dumas, E.; Guerlin, A.; Miomandre, F.; Clavier, G.; Bertin, D.; Gigmes, D. Cationic iridium complex for the design of soft

salt-based phosphorescent OLEDs and color-tunable light-emitting electrochemical cells. *Org. Electron.* 2011, 12, 1683–1694 b) Nasr, G.; Guerlin, A.; Dumur, F.; Beouch, L.; Dumas, E.; Clavier, G.; Miomandre, F.; Goubard, F.; Gigmes, D.; Bertin, D.; Wantz, G.; Mayer, C.R. Iridium (III) soft salts from dinuclear cationic and mononuclear anionic complexes for OLED devices. *Chem. Commun.* 2011, 47, 10698–10700.

[11] a) Murawski, C.; Leo, K.; Gather, M.C. Efficiency roll-off in organic light-emitting diodes. *Adv. Mater.* 2013, 25, 6801–6827 b) Wehrmeister, S.; Jäger, L.; Wehlus, L.; Rausch, A.F.; Reusch, T.C.G.; Schmidt, T.D.; Brütting, W. Combined electrical and optical analysis of the efficiency roll-off in phosphorescent organic light-emitting diodes, *Phys. Rev. Appl.* 2015, 3, 024008-1–024008-10 c) Zhang, T.; Zhao, B.; Chu, B.; Li, W.; Su, Z.; Yan, X.; Liu, C.; Wu, H.; Gao, Y.; Jin, F.; Hou, F. Simple structured hybrid WOLEDs based on incomplete energy transfer mechanism: from blue exciplex to orange dopant, *Sci. Report* 2015, 5, 10234-1–10234-8.

[12] a) Baldo, M.A.; Adachi, C.; Forrest, S.R. Transient analysis of organic electrophosphorescence. II. Transient analysis of triplet-triplet annihilation. *Phys. Rev. B* 2000, 62, 10967–10977 b) Reineke, S.; Walzer, K.; Leo, K. Triplet-exciton quenching in organic phosphorescent light-emitting diodes with Ir-based emitters. *Phys. Rev. B* 2007, 75, 125328-1–125328-13 c) Murawski, C.; Liehm, P.; Leo, K.; Gather, M.C. Influence of cavity thickness and emitter orientation on the efficiency roll-off of phosphorescent organic light-emitting diodes. *Adv. Mater.* 2014, 24, 1117–1124.

[13] a) Kalinowski, J.; Stampor, W.; Mezyk, J.; Cocchi, M.; Virgili, D.; Fattori, V.; Di Marco, P. Quenching effects in organic electrophosphorescence. *Phys. Rev. B* 2002, 66, 235321-1–235321-15 b) Cocchi, M.; Fattori, V.; Virgili, D.; Sabatini, C.; Marco, P.D. Highly efficient organic electrophosphorescent light-emitting diodes with a reduced quantum efficiency roll off at large current densities. *Appl. Phys. Lett.* 2004, 84, 1052–1054 c) Cocchi, M.; Kalinowski, J. Single-dopant organic white electrophosphorescent diodes with very high efficiency and its reduced current density roll-off. *Appl. Phys. Lett.* 2007, 90, 163508-1–163508-3.

[14] a) Kang, J.-W.; Lee, S.-H.; Park, H.-D.; Jeong, W.-I.; Yoo, K.-M.; Park, Y.-S.; Kim, J.-J. Low roll-off of efficiency at high current density in phosphorescent organic light emitting diodes. *Appl. Phys. Lett.* 2007, 90, 223508-1–223508-3 b) Gao, Z.; Wang, F.; Guo, K.; Wang, H.; Wei, B.; Xu, B. Carrier transfer and luminescence characteristics of concentration-dependent phosphorescent $Ir(ppy)_3$ doped CBP film. *Opt Laser Technol.* 2014, 56, 20–24.

[15] a) He, G.; Schneider, O.; Qin, D.; Zhou, X.; Pfeiffer, M.; Leo, K. Very high-efficiency and low voltage phosphorescent organic light-emitting diodes based on a *p-i-n* junction. *J. Appl. Phys.* 2004, 95, 5773–5777 b) Kalinowski, J.; Stampor, W.; Cocchi, M.; Virgili, D.; Fattori, V.; Marco, P.D. Triplet energy exchange between fluorescent and phosphorescent organic molecules in a solid state matrix. *Chem. Phys.* 2004, 297, 39–48 c) Zang, F.X.; Sum, T.C.; Huan, A.C.H.; Li, T.L.; Li, W.L.; Zhu, F. Reduced efficiency roll-off in phosphorescent organic light emitting diodes at ultrahigh current densities by suppression of triplet-polaron quenching. *Appl. Phys. Lett.* 2008, 93, 023309-1–023309-3.

[16] a) Baldo, M.A.; Adachi, C.; Forrest, S.R. Transient analysis of organic electrophosphorescence. II. Transient analysis of triplet-triplet annihilation. *Phys. Rev. B* 2000, 62, 10967–10977 b) Giebink, N.C.; Forrest, S.R. Quantum efficiency roll-

off at high brightness in fluorescent and phosphorescent organic light emitting diodes. *Phys. Rev. B* 2008, 77, 235215-1–235215-9.

[17] Park, Y.-S.; Jeong, W.-I.; Kim, J.-J. Energy transfer from exciplexes to dopants and its effect on efficiency of organic light-emitting diodes. *J. Appl. Phys.* 2011, 110, 124519-1–124519-6.

[18] a) Reineke, S.; Baldo, M.A. Recent progress in the understanding of exciton dynamics within phosphorescent OLEDs. *Phys Status Solidi A* 2012, 209, 2341–2353 b) Reineke, S.; Schwartz, G.; Walzer, K.; Falke, M.; Leo, K. Highly phosphorescent organic mixed films: The effect of aggregation on triplet-triplet annihilation. *Appl. Phys. Lett.* 2009, 94, 163305-1–163305-3 c) Liu, S.W.; Divayana, Y.; Abiyasa, A.P.; Tan, S.T.; Demir, H.V.; Sun, X.W. On the triplet distribution and its effect on an improved phosphorescent organic light-emitting diode. *Appl. Phys. Lett.* 2012, 101, 093301-1–093301-4.

[19] a) Li, J.; Zhang, T.; Liang, Y.; Yang, R. Solution-processible carbazole dendrimers as host materials for highly efficient phosphorescent organic light-emitting diodes. *Adv. Funct. Mater.* 2013, 23, 619–628 b) Tao, Y.; Yang, C.; Qin, J. Organic host materials for phosphorescent organic light-emitting diodes. *Chem. Soc. Rev.* 2011, 40, 2943–2970.

[20] a) Adachi, C.; Kwong, R.C.; Djurovich, P.; Adamovich, V.; Baldo, M.A.; Thompson, M.E.; Forrest, S.R. Endothermic energy transfer: A mechanism for generating very efficient high-energy phosphorescent emission in organic materials. *Appl. Phys. Lett.* 2001, 79, 2082–2084 b) Deng, L.; Wang, X.; Zhang, Z.; Li, J. Durene-decorated CBP derivatives as phosphorescent hosts and exciton-blocking materials for efficient blue OLEDs. *J. Mater. Chem.* 2012, 22, 19700–19708 c) Tokito, S.; Lijima, T.; Suzuri, Y.; Kita, H.; Tsuzuki, T.; Sato, F. Confinement of triplet energy on phosphorescent molecules for highly-efficient organic blue-light-emitting devices. *Appl. Phys. Lett.* 2003, 83, 569–571.

[21] a) Chaskar, A.; Chen, H.-F.; Wong, K.-T. Bipolar host materials: A chemical approach for highly efficient electrophosphorescent devices. *Adv. Mater.* 2011, 23, 3876–3895 b) Dumur, F.; Goubard, F. Triphenylamines and 1,3,4-oxadiazoles: a versatile combination for controlling the charge balance in organic electronics. *New J. Chem.* 2014, 38, 2204–2224 c) Xu, Q.-L.; Liang, X.; Zhang, S.; Jing, Y.-M.; Liu, X.; Lu, G.-Z.; Zheng, Y.-X.; Zuo, J.-L. Efficient OLEDs with low efficiency roll-off using iridium complexes possessing good electron mobility. *J. Mater. Chem.* C 2015, 3, 3694-3701.

[22] Yang, X.; Zhuang, S.; Qiao, X.; Mu, G.; Wang, L.; Chen, J.; Ma, D. High efficiency blue phosphorescent organic light-emitting diodes with a multiple quantum well structure for reduced efficiency roll-off. *Opt. Express* 2012, 20, 24411–24417.

[23] a) Agata, Y.; Shimizu, H.; Kido, J. Syntheses and properties of novel quarterphenylene-based materials for blue organic light-emitting devices. *Chem. Lett.* 2007, 36, 316–317 b) Tsai, M.-H.; Lin, H.-W.; Su, H.-C.; Ke, T. H.; Wu, C.-C.; Fang, F.- C.; Liao, Y.-L.; Wong, K.-T.; Wu, C.-I. Highly efficient organic blue electrophosphorescent devices based on 3,6-*bis*(triphenylsilyl)carbazole as the host material. *Adv. Mater.* 2006, 18, 1216–1220 c) Tsai, M.-H.; Hong, Y.-H.; Chang, C.-H.; Su, H.-C.; Wu, C.-C.; Matoliukstyte, A.; Simokaitiene, J.; Grigalevicius, S.; Grazulevicius, J.V.; Hsu, C.-P. 3-(9-Carbazolyl)carbazoles and 3,6-di(9-carbazolyl)carbazoles as effective host materials for efficient blue organic electrophosphorescence. *Adv. Mater.* 2007, 19, 862–866 d)

Yeh, S.-J.; Wu, M.- F.; Chen, C.-T.; Song, Y.-H.; Chi, Y.; Ho, M.-H.; Hsu, S.-F.; Chen, C.H. New dopant and host materials for blue-light-emitting phosphorescent organic electroluminescent devices. *Adv. Mater.* 2005, 17, 285–289.

[24] Holmes, R.J.; D'Andrade, B.W.; Forrest, S.R.; Ren, X.; Thompson, M.E. Efficient, deep-blue organic electro-phosphorescence by guest charge trapping. *Appl. Phys. Lett.* 2003, 83, 3818–3820.

[25] Wang, Z.B.; Helander, M.G.; Qiu, J.; Puzzo, D.P.; Greiner, M.T.; Liu, Z.W.; Lu, Z.H. Highly simplified phosphorescent organic light emitting diode with > 20% external quantum efficiency at > 10,000 cd/m^2. *Appl. Phys. Lett.* 2011, 98, 073310-1–073310-3.

[26] Jeon, S.O.; Jang, S.E.; Son, H.S.; Lee, J.Y. External quantum efficiency above 20% in deep blue phosphorescent organic light-emitting diodes. *Adv. Mater.* 2011, 23, 1436–1441.

[27] a) Chang, C.-F.; Cheng, Y.-M.; Chi, Y.; Chiu, Y.-C.; Lin, C.-C.; Lee, G.-H.; Chou, P.-T.; Chen, C.-C.; Chang, C.-H.; Wu, C.-C. Highly efficient blue-emitting iridium (III) carbene complexes and phosphorescent OLEDs. *Angew Chem. Int. Ed* 2008, 47, 4542–4545 b) Chiu, Y.-C.; Hung, J.-Y.; Chi, Y.; Chen, C.-C.; Chang, C.-H.; Wu, C.-C.; Cheng, Y.-M.; Yu, Y.-C.; Lee, G.-H.; Chou, P.-T. En route to high external quantum efficiency (~12%), organic true-blue-light-emitting diodes employing novel design of iridium (III) phosphors. *Adv. Mater.* 2009, 10, 2221–2225 c) Sajoto, T.; Djurovich, P.I.; Tamayo, A.; Yousufuddin, M.; Bau, R.; Thompson, M.E.; Holmes, R.J.; Forrest, S.R. Blue and near-UV phosphorescence from iridium complexes with cyclometalated pyrazolyl or *N*-heterocyclic carbene ligands. *Inorg Chem* 2005, 44, 7992–8003 d) Erk, P.; Bold, M.; Egen, M.; Fuchs, E.; Geßner, T.; Kahle, K.; Lennartz, C.; Molt, O.; Nord, S.; Reichelt, H.; Schildknecht, C.; Johannes, H.-H.; Kowalsky, W. Efficient deep blue triplet emitters for OLEDs. *SID Digest* 2006, 37, 131–134 e) Kim, S.H.; Jang, J.; Lee, S.J.; Lee, J.Y. Deep blue phosphorescent organic light-emitting diodes using a Si based wide bandgap host and an Ir dopant with electron withdrawing substituents. *Thin Solid Films* 2008, 517, 722–726 f) Yook, K.S.; Jeon, S.O.; Joo, C.W.; Lee, J.Y. High efficiency deep blue phosphorescent organic light-emitting diodes. *Org. Electron* 2009, 10, 170–173 g) Holmes, R.J.; Forrest, S.R.; Sajoto, T.; Tamayo, A.; Djurovich, P.I.; Thompson, M.E.; Brooks, J.; Tung, H.J.; D'Andrade, B.W.; Weaver, M.S.; Kwong, R.C.; Brown, J.J. Saturated deep blue organic electrophosphorescence using a fluorine-free emitter. *Appl. Phys. Lett.* 2005, 87, 243507-1–243507-3 h) Jeon, S.O.; Yook, K.S.; Joo, C.W.; Lee, J.Y.; High-efficiency deep-blue-phosphorescent organic light-emitting diodes using a phosphine oxide and a phosphine sulfide high-triplet-energy host material with bipolar charge-transport properties. *Adv. Mater.* 2010, 22, 1872–1876 i) Jeon, S.O.; Yook, K.S.; Joo, C.W.; Lee, J.Y. Phenylcarbazole-based phosphine oxide host materials for high efficiency in deep blue phosphorescent organic light-emitting diodes. *Adv. Funct. Mater.* 2009, 19, 3644–3649 j) Chiu, Y.-C.; Chi, Y.; Hung, J.-Y.; Cheng, Y.-M.; Yu, Y.-C.; Chung, M.-W.; Lee, G.-H.; Chou, P.-T.; Chen, C.-C.; Wu, C.-C.; Hsieh, H.-Y. Blue to true-blue phosphorescent Ir^{III} complexes bearing a nonconjugated ancillary phosphine chelate: strategic synthesis, photophysics, and device integration. *ACS Appl. Mater. Interfaces* 2009, 1, 433–442 k) Su, S.-J.; Chiba, T.; Takeda, T.; Kido, J. Pyridine-containing triphenylbenzene derivatives with high electron mobility for highly efficient phosphorescent OLEDs. *Adv. Mater.* 2008, 20, 2125–2130 l) Sasabe, H.; Gonmori, E.; Chiba, T.; Li, Y.-J.; Tanaka, D., Su, S.-J.;

Takeda, T.; Pu, Y.-J.; Nakayama K., Kido, J. Wide-energy-gap electron-transport materials containing 3,5-dipyridylphenyl moieties for an ultra high efficiency blue organic light-emitting device. *Chem. Mater.* 2008, 20, 5951–5953 m) Chopra, N.; Lee, J.; Zheng, Y.; Eom, S.-H.; Xue, J.; So, F. High efficiency blue phosphorescent organic light-emitting device. *Appl. Phys. Lett.* 2008, 93, 143307-1–143307-3.

[28] Padmaperuma, A.B.; Sapochak, L.S.; Burrows, P.E. New charge transporting host material for short wavelength organic electrophosphorescence: 2,7-*bis*(diphenylphosphine oxide)-9,9-dimethylfluorene. *Chem. Mater.* 2006, 18, 2389–2396.

[29] a) Yang, C.-H.; Liao, S.-H.; Sun, Y.-K.; Chuang, Y.-Y.; Wang, T.-L.; Shieh, Y.-T.; Lin, W.-C. Optimization of multiple electron donor and acceptor in carbazole-triphenylamine-based molecules for application of dye-sensitized solar cells. *J. Phys. Chem. C* 2010, 114, 21786–21794 b) Bui, T.-T.; Beouch, L.; Sallenave, X.; Goubard, F. Carbazol-*N*-yl and diphenylamino end-capped triphenylamine-based molecular glasses: synthesis, thermal, and optical properties. *Tetrahedron Lett.* 2013, 54, 4277–4280.

[30] Su, S.-J.; Sasabe, H.; Takeda, T.; Kido, J. Pyridine-containing bipolar host materials for highly efficient blue phosphorescent OLEDs. *Chem. Mater.* 2008, 20, 1691–1693.

[31] Su, S.-J.; Cai, C.; Kido, J. RGB phosphorescent organic light-emitting diodes by using host materials with heterocyclic cores: effect of nitrogen atom orientations. *Chem. Mater.* 2011, 23, 274–284.

[32] Yang, X.; Huang, H.; Pan, B.; Zhuang, S.; Aldred, M.P.; Wang, L.; Chen, J.; Ma, D. Novel electron-type host material for unilateral homogeneous phosphorescent organic light-emitting diodes with low efficiency roll-off. *J. Mater. Chem.* 2012, 22, 23129–23135.

[33] Su, S.-J.; Cai, C.; Kido, J. Three-carbazole-armed host materials with various cores for RGB phosphorescent organic light-emitting diodes. *J. Mater. Chem.* 2012, 22, 3447–3456.

[34] Jeon, S.-O.; Yook, K.S.; Joo, C.W.; Lee, J.Y.; Ko, K.-Y.; Park, J.-Y.; Baek, Y.G. 100% internal quantum efficiency and stable efficiency roll-off in phosphorescent light-emitting diodes using a high triplet energy hole transport material. *Appl. Phys. Lett.* 2008, 93, 063306-1–063306-3.

[35] Cho, Y.J.; Lee, J.Y. Low driving voltage, high quantum efficiency, high power efficiency, and little efficiency roll-off in red, green, and deep-blue phosphorescent organic light-emitting diodes using a high-triplet-energy hole transport material. *Adv. Mater.* 2011, 23, 4568–4572.

[36] Mondal, E.; Hung, W.-Y.; Chen, Y.-H.; Cheng, M.-H.; Wong, K.-T. Molecular topology tuning of bipolar host materials composed of fluorene-bridged benzimidazole and carbazole for highly efficient electrophosphorescence. *Chem. Eur. J.* 2013, 19, 10563–10572.

[37] Li, W.; Qiao, J.; Duan, L.; Wang, L.; Qiu, Y. Novel fluorene/carbazole hybrids with steric bulk as host materials for blue organic electrophosphorescent devices. *Tetrahedron* 2007, 63, 10161–10168.

[38] a) Ikai, M.; Ishikawa, F.; Aratani, N.; Osuka, A.; Kawabata, S.; Kajioka, T.; Takeuchi, H.; Fujikawa, H.; Taga, Y. Enhancement of external quantum efficiency of red phosphorescent organic light-emitting devices with facially encumbered and bulky Pt^{II} porphyrin complexes. Adv. Funct. Mater. 2006, 16, 515–519 b) Zhu, M.; Yang, C. Blue

fluorescent emitters: design tactics and applications in organic light-emitting diodes. *Chem. Soc. Rev.* 2013, 42, 4963–4976.

[39] Tsai, M.-H.; Ke, T.-H.; Lin, H.-W.; Wu, C.-C.; Chiu, S.-F.; Fang, F.-C.; Liao, Y.-L.; Wong, K.-T.; Chen, Y.-H.; Wu, C.-I. Triphenylsilyl- and trityl-substituted carbazole-based host materials for blue electrophosphorescence. *ACS Appl. Mater. Interfaces* 2009, 1, 567–574.

[40] Chen, H.; Jiang, Z.-Q.; Gao, C.-H.; Xu, M.-F.; Dong, S.-C.; Cui, L.-S.; Ji, S.-J.; Liao, L.-S. Silicon-based material with spiro-annulated fluorene/triphenylamine as host and exciton-blocking layer for blue electrophosphorescent devices. *Chem. Eur. J.* 2013, 19, 11791–11797.

[41] a) Ren, X.; Li, J.; Holmes, R.J.; Djurovich, P.I.; Forrest, S.R.; Thompson, M.E. Ultrahigh energy gap hosts in deep blue organic electrophosphorescent devices. *Chem. Mater.* 2004, 16, 4743–4747 b) Lin, J.-J.; Liao, W.-S.; Huang, H.-J.; Wu, F.-I.; Cheng, C.-H. A highly efficient host/dopant combination for blue organic electrophosphorescence devices. *Adv. Funct. Mater.* 2008, 18, 485–491 c) Lin, M.-S.; Chi, L.-C.; Chang, H.-W.; Huang, Y.-H.; Tien, K.-C.; Chen, C.-C.; Chang, C.-H.; Wu, C.-C.; Chaskar, A.; Chou, S.-H.; Ting, H.-C.; Wong, K.-T.; Liu, Y.-H.; Chi, Y. A diarylborane-substituted carbazole as a universal bipolar host material for highly efficient electrophosphorescence devices. *J. Mater. Chem.* 2012, 22, 870–876 d) Kim, Y.H.; Jeong, H.C.; Kim, S.H.; Yang, K.; Kwon, S.K. High-purity-blue and high-efficiency electroluminescent devices based on anthracene. *Adv. Funct. Mater.* 2005, 15, 1799–1805 e) Kang, J.-W.; Lee, D.-S.; Park, H.-D.; Kim, J.W.; Jeong, W.-I.; Park, Y.-S.; Lee, S.-H.; Go, K.; Lee, J.-S.; Kim, J.-J. A host material containing tetraphenylsilane for phosphorescent OLEDs with high efficiency and operational stability. *Org. Electron* 2008, 9, 452–460 f) Yeh, S.-J.; Wu, M.-F.; Chen, C.-T.; Song, Y.-H.; Chi, Y.; Ho, M.-H.; Hsu, S.-F.; Chen, C.H. New dopant and host materials for blue-light-emitting phosphorescent organic electroluminescent devices. *Adv. Mater.* 2005, 17, 285–289 g) Han, W.-S.; Son, H.-J.; Wee, K.-R.; Min, K.-T.; Kwon, S.; Suh, I.-H.; Choi, S.-H.; Jung, D.H.; Kang, S.O. Silicon-based blue phosphorescence host materials: structure and photophysical property relationship on methyl/phenylsilanes adorned with 4-(*N*-carbazolyl)phenyl groups and optimization of their electroluminescence by peripheral 4-(*N*-carbazolyl)phenyl numbers. *J. Phys. Chem. C* 2009, 113, 19686–19693 h) Jiang, Z.; Chen, Y.; Fan, C.; Yang, C.; Wang, Q.; Tao, Y.; Zhang, Z.; Qin, J.; Ma, D. Bridged triphenylamines as novel host materials for highly efficient blue and green phosphorescent OLEDs. *Chem. Commun.* 2009), 3398–3400 i) Leung, M.-K.; Yang, W.-H.; Chuang, C.-N.; Lee, J.-H.; Lin, C.-F.; Wei, M.-K.; Liu, Y.-H. 1,3,4-Oxadiazole containing silanes as novel hosts for blue phosphorescent organic light emitting diodes. *Org. Lett.* 2012, 14, 4986–4989 j) Gong, S.; Chen, Y.; Yang, C.; Zhong, C.; Qin, J.; Ma, D. De novo design of silicon-bridged molecule towards a bipolar host: all-phosphor white organic light-emitting devices exhibiting high efficiency and low efficiency roll-off. *Adv. Mater.* 2010, 22, 5370–5373.

[42] Jang, S.E.; Jeon, S.O.; Cho, Y.J.; Yook, K.S.; Lee, J.Y. Stable efficiency roll-off in red phosphorescent organic light-emitting diodes using a spirofluorene–benzofluorene based carbazole type host material. *J. Lumin.* 2010, 130, 2184–2187.

[43] Dong, S.-C.; Gao, C.-H.; Yuan, X.-D.; Cui, L.-S.; Jiang, Z.-Q.; Lee, S.-T.; Liao, L.-S. Novel dibenzothiophene based host materials incorporating spirobifluorene for high-

efficiency white phosphorescent organic light-emitting diodes. *Org. Electron* 2013, 14, 902–908.

[44] Kwak, J.; Lyu, Y.-Y.; Lee, H.; Choi, B.; Char, K.; Lee, C. New carbazole-based host material for low-voltage and highly efficient red phosphorescent organic light-emitting diodes. *J. Mater. Chem.* 2012, 22, 6351–6355.

[45] a) Adachi, C.; Baldo, M.A.; Forrest, S.R.; Lamansky, S.; Thompson, M.E.; Kwong, R.C. High-efficiency red electrophosphorescence devices. *Appl. Phys. Lett.* 2001, 78, 1622–1624 b) Wong, K.-T.; Chen, Y.-M.; Lin, Y.-T.; Su, H.-C.; Wu, C.-C. Nonconjugated hybrid of carbazole and fluorene: a novel host material for highly efficient green and red phosphorescent OLEDs. *Org. Lett.* 2005, 7, 5361–5364 c) Wong, W.-Y.; Ho, C.-L.; Gao, Z.-Q.; Mi, B.-X.; Chen, C.-H.; Cheah, K.-W.; Lin Z. Soluble poly(*p*-phenylenevinylene)s through ring-opening metathesis polymerization. *Angew Chem.* 2006, 118, 7961–7964 d) Shih, P.-I.; Chiang, C.-L.; Dixit, A.K.; Chen, C.-K.; Yuan, M.-C.; Lee, R.-Y.; Chen, C.-T.; Diau, E.W.-G.; Shu, C.-F. Novel carbazole/fluorene hybrids: host materials for blue phosphorescent OLEDs. *Org. Lett.* 2006, 8, 2799–2802.

[46] Ting, H.-C.; Chen, Y.-M.; You, H.-W.; Hung, W.-Y.; Lin, S.-H.; Chaskar, A.; Chou, S.-H.; Chi, Y.; Liu, R.-H.; Wong, K.-T. Indolo[3,2-b]carbazole/benzimidazole hybrid bipolar host materials for highly efficient red, yellow, and green phosphorescent organic light emitting diodes. *J. Mater. Chem.* 2012, 22, 8399–8407.

[47] a) Tao, Y.; Wang, Q.; Yang, C.; Wang, Q.; Zhang, Z.; Zou, T.; Qin, J.; Ma, D. A simple carbazole/oxadiazole hybrid molecule: an excellent bipolar host for green and red phosphorescent OLEDs. *Angew Chem. Int. Ed* 2008, 47, 8104–8107 b) Lyu, Y.-Y.; Kwak, J.; Jeon, W.S.; Byun,Y.; Lee, H.S.; Kim, D.; Lee, C.; Char, K. Highly efficient red phosphorescent OLEDs based on non-conjugated silicon-cored spirobifluorene derivative doped with Ir-complexes. *Adv. Funct. Mater.* 2009 19, 420–427 c) Chien, C.-H.; Hsu, F.-M.; Shu, C.-F.; Chi, Y. Efficient red electrophosphorescence from a fluorene-based bipolar host material. *Org. Electron* 2009, 10, 871–876 d) Tao, Y.; Wang, Q.; Ao, L.; Zhong, C.; Qin, J.; Yang, C.; Ma, D. Molecular design of host materials based on triphenylamine/oxadiazole hybrids for excellent deep-red phosphorescent organic light-emitting diodes. *J. Mater. Chem.* 2010, 20, 1759–1765.

[48] a) O'Brien, D.F.; Burrows, P.E.; Forrest, S.R.; Konne, B.E.; Loy, D.E.; Thompson, M.E. Hole transporting materials with high glass transition temperatures for use in organic light-emitting devices. *Adv. Mater.* 1998, 10, 1108–1112 b) Katsuma, A.K.; Shirota, Y. A novel class of π-electron dendrimers for thermally and morphologically stable amorphous molecular materials. *Adv. Mater.* 1998, 10, 223–226 c) Ko, C.W.; Tao, T. 9,9-*Bis*{4-[di-(*p*-biphenyl)aminophenyl]}fluorene: a high T_g and efficient hole-transporting material for electroluminescent devices. *Synth Met* 2002, 126, 37–41 d) Salbeck, J.; Bauer, J.; Weissörtel, F. Spiro linked compounds for use as active materials in organic light emitting diodes. *Macromol. Symp.* 1998, 125, 121–132 e) Xiao, H.; Leng, B.; Tian, H. Hole transport triphenylamine–spirosilabifluorene alternating copolymer: synthesis and optical, electrochemical and electroluminescent properties. *Polymer* 2005, 46, 5707–5713 f) Xiao, H.; Shen, H.; Lin, Y.; Su, J.; Tian, H. Spirosilabifluorene linked bistriphenylamine: Synthesis and application in hole transporting and two-photon fluorescent imaging. *Dyes Pigments* 2007, 73, 224–229 g) Katsis, D.; Geng, Y.H.; Ou, J.J.; Culligan, S.W.; Trajkovska, A.; Chen, S.H.; Rothberg,

L.J. Spiro-linked ter-, penta-, and heptafluorenes as novel amorphous materials for blue light emission. *Chem. Mater.* 2002, 14, 1332–1339.

[49] Kim, J.-Y.; Lee, C.-W.; Jang, J.G.; Gong, M.-S. Orange phosphorescent organic light-emitting diodes using new spiro[benzoanthracene-fluorene]-type host materials. *Dyes Pigments* 2012, 94, 304–313.

[50] Park, J.S.; Jeon, W.S.; Yu, J.H.; Pode, R.; Kwon, J.H. Efficiency optimization of green phosphorescent organic light-emitting device. *Thin Solid Films* 2011, 519, 3259–3263.

[51] Deng, L.; Li, J.; Wang, G.-X.; Wu, L.-Z. Simple bipolar host materials incorporating CN group for highly efficient blue electrophosphorescence with slow efficiency roll-off. *J. Mater. Chem. C* 2013, 1, 8140–8145.

[52] a) Zhang, T.; Liang, Y.; Cheng, J.; Li, J. A CBP derivative as bipolar host for performance enhancement in phosphorescent organic light-emitting diodes. *J. Mater. Chem. C* 2013, 1, 757–764 b) Hung, W.Y.; Tsai, T.C.; Ku, S.Y.; Chi, L.C.; Wong, K.T. An ambipolar host material provides highly efficient saturated red PhOLEDs possessing simple device structures. *Phys. Chem. Chem. Phys.* 2008, 10, 5822–5825.

[53] Bin, J.K.; Cho, N.S.; Hong, J.I. New host material for high-performance blue phosphorescent organic electroluminescent devices. *Adv. Mater.* 2012, 24, 2911–2915.

[54] Liu, H.; Cheng, G.; Hu, D.; Shen, F.; Lv, Y.; Sun, G.; Yang, B.; Lu, P.; Ma, Y. A highly efficient, blue-phosphorescent device based on a wide-bandgap host/FIrpic: rational design of the carbazole and phosphine oxide moieties on tetraphenylsilane. *Adv. Funct. Mater.* 2012, 22, 2830–2836.

[55] Lin, M.S.; Yang, S.J.; Chang, H.W.; Huang, Y.H.; Tsai, Y.T.; Wu, C.C.; Chou, S.H.; Mondal, E.; Wong, K.T. Incorporation of a CN group into mCP: a new bipolar host material for highly efficient blue and white electrophosphorescent devices. *J. Mater. Chem.* 2012, 22, 16114–16120.

[56] a) Lee, J.; Lee, J.-I.; Lee, J.Y.; Chu, H.Y. Stable efficiency roll-off in blue phosphorescent organic light-emitting diodes by host layer engineering. *Org. Electron* 2009, 10, 1529–1533 b) Yook, K.S.; Jeon, S.O.; Joo, C.W.; Lee, J.Y.; Kim, M.S.; Choi, H.S.; Lee, S.J.; Han, C.-W.; Tak, Y.H. Highly efficient pure white phosphorescent organic light-emitting diodes using a deep blue phosphorescent emitting material. *Org. Electron* 2009, 10, 681–685 c) Chopra, N.; Swensen, J.S.; Polikarpov, E.; Cosimbescu, L.; So, F.; Padmaperuma, A.B. High efficiency and low roll-off blue phosphorescent organic light-emitting devices using mixed host architecture. *Appl. Phys. Lett.* 2010, 97, 033304-1–033304-3.

[57] Park, Y.-S.; Lee, S.; Kim, K.-H.; Kim, S.-Y.; Lee, J.-H.; Kim, J.-J. Exciplex-forming co-host for organic light-emitting diodes with ultimate efficiency. *Adv. Funct. Mater.* 2013, 23, 4914–4920.

[58] Jang, S.E.; Lee, J.Y. High power efficiency in blue phosphorescent organic light-emitting diodes using a spirobifluorene phosphine oxide compound. *Mol. Cryst. Liq. Cryst.* 2011, 551, 206–211.

[59] Suh, M.C.; Shin, H.Y.; Cha, S.J. Highly efficient green phosphorescent organic light emitting diodes with simple structure. *Org. Electron.* 2013, 14, 2198–2203.

[60] Chen, J.; Shi, C.; Fu, Q.; Zhao, F.; Hu, Y.; Feng, Y.; Ma, D. Solution-processable small molecules as efficient universal bipolar host for blue, green and red phosphorescent inverted OLEDs. *J. Mater. Chem.* 2012, 22, 5164–5170.

[61] a) Yook, K.S.; Jang, S.E.; Jeon, S.O.; Lee, J.Y. Fabrication and efficiency improvement of soluble blue phosphorescent organic light-emitting diodes using a multilayer structure based on an alcohol-soluble blue phosphorescent emitting layer. *Adv. Mater.* 2010, 22, 4479–4483 b) Lee, H.; Park, I.; Kwak, J.; Yoon, D.Y.; Lee, C. Improvement of electron injection in inverted bottom-emission blue phosphorescent organic light emitting diodes using zinc oxide nanoparticles. *Appl. Phys. Lett.* 2010, 96, 153306-1–153306-3.

[62] Kim, S.H.; Jang, J.; Yook, K.S.; Lee, J.Y. Stable efficiency roll-off in phosphorescent organic light-emitting diodes. *Appl. Phys. Lett.* 2008, 92, 023513-1–02351363.

[63] Lee, J.; Lee, J.-I.; Lee, J.Y.; Chu, H.Y. Enhanced efficiency and reduced roll-off in blue and white phosphorescent organic light-emitting diodes with a mixed host structure. *Appl. Phys. Lett.* 2009, 94, 193305-1–193305-3.

[64] Lee, J.; Lee, J.-I.; Lee, J.-W.; Chu, H.Y. Effects of charge balance on device performances in deep blue phosphorescent organic light-emitting diodes. *Org. Electron.* 2010, 11, 1159–1164.

[65] Lee, J.; Lee, J.-I.; Lee, J.Y.; Chu, H.Y. Stable efficiency roll-off in blue phosphorescent organic light-emitting diodes by host layer engineering. *Org. Electron.* 2009, 10, 1529–1533.

[66] Chen, Y.; Chen, J.; Zhao, Y.; Ma, D. High efficiency blue phosphorescent organic light-emitting diode based on blend of hole- and electron-transporting materials as a co-host. *Appl. Phys. Lett.* 2012, 100, 213301-1–213301-4.

[67] Wang, J.; Liu, J.; Huang, S.; Wu, X.; Shi, X.; Chen, C.; Ye, Z.; Lu, J.; Su, Y.; He, G.; Zheng, Y. High efficiency green phosphorescent organic light-emitting diodes with a low roll-off at high brightness. *Org. Electron.* 2013, 14, 2854–2858.

[68] Lee, S.; Kim, K.-H.; Limbach, D.; Park, Y.-S.; Kim, J.-J. Low roll-off and high efficiency orange organic light emitting diodes with controlled co-doping of green and red phosphorescent dopants in an exciplex forming co-host. *Adv. Funct. Mater.* 2013, 23, 4105–4110.

[69] Lee, S.; Limbach, D.; Kim, K.-H.; Yoo, S.-J.; Park, Y.-S.; Kim, J.-J. High efficiency and non-color-changing orange organic light emitting diodes with red and green emitting layers. *Org. Electron.* 2013, 14, 1856–1860.

[70] Song, M.; Park, J.S.; Yoon, M.; Kim, A.J.; Kim, Y.I., Gal, Y.-S.; Lee, J.W.; Jin, S.-H. Solution-processed red iridium complexes based on carbazole-phenylquinoline main ligand: Synthesis, properties and their applications in phosphorescent organic light-emitting diodes. *J. Organomet. Chem.* 2011, 696, 2122–2128.

[71] Pina, J.; Seixas de Melo, J.; Burrows, H.D.; Monkman, A.P.; Navaratnam, S. On the triplet state of poly(*N*-vinylcarbazole). *Chem. Phys. Lett.* 2004, 400, 441–445.

[72] Song, M.; Park, J.S.; Gal, Y.-S.; Kang, S.; Lee, J.Y.; Lee, J.W.; Jin, S.-H. Homologous series of phenylquinoline-carbazole main ligand based on red-emitting iridium (III) complexes for phosphorescent organic light-emitting diodes. *J. Phys. Chem. C* 2012, 116, 7526–7533.

[73] Jou, J.-H.; Chen, C.-H.; Tseng, J.-R.; Peng, S.-H.; Chen, P.-W.; Chiang, C.-I.; Jou, Y.-C.; Hong, J.H.; Wang, C.-C.; Chen, C.-C.; Tung, F.-C.; Chen, S.-H.; Wang, Y.-S.; Chin, C.-L. Using light-emitting dyes as a co-host to markedly improve efficiency roll-off in phosphorescent yellow organic light emitting diodes. *J. Mater. Chem. C* 2013, 1, 394–400.

[74] a) Chang, Y.-L.; Wang, Z.B.; Helander, M.G.; Qiu, J.; Puzzo, D.P.; Lu, Z.H. Enhancing the efficiency of simplified red phosphorescent organic light emitting diodes by exciton harvesting. *Org. Electron.* 2012, 13, 925–931; b) Yoo, S. I.; Yoon, J.A.; Kim, N.H.; Kim, J.W.; Kang, J.S.; Moon, C.-B.; Kim, W.Y. Improvement of efficiency roll-off in blue phosphorescence OLED using double dopants emissive layer. *J. Lumin.* 2015, 160, 346–350.

[75] Wang, Q.; Ding, J.; Ma, D.; Cheng, Y.; Wang, L. Highly efficient single-emitting-layer white organic light-emitting diodes with reduced efficiency roll-off. *Appl. Phys. Lett.* 2009, 94, 103503-1–103503-3.

[76] Gong, S.; Chen, Y.; Luo, J.; Yang, C.; Zhong, C.; Qin, J.; Ma, D. Bipolar tetraarylsilanes as universal hosts for blue, green, orange, and white electrophosphorescence with high efficiency and low efficiency roll-off. *Adv. Funct. Mater.* 2011, 21, 1168–1178.

[77] Guo, R.; Hu, D.; Yue, S.; Zhang, Z.; Duan, Y.; Lu, P.; Ma, Y.; Zhao, Y.; Liu S. Novel tetraarylsilan-centred compounds as single host for white organic light-emitting diodes with high efficiency and low roll-off. *J. Phys. D. Appl. Phys.* 2013, 46, 265101-1–265101-6.

[78] Fleetham, T.; Wang, Z.; Li, J. Efficient deep blue electrophosphorescent devices based on platinum(II) *bis*(*n*-methyl-imidazolyl)benzene chloride. *Org. Electron.* 2012, 13, 1430-1435.

[79] a) Adamovich, V.; Brooks, J.; Tamayo, A.; Alexander, A.M.; Djurovich, P.I.; D'Andrade, B.W.; Adachi, C.; Forrest, S.R.; Thompson, M.E. High efficiency single dopant white electrophosphorescent light emitting diodes. *New J. Chem.* 2002, 26, 1171–1178; b) Williams, E.L.; Haavisto, K.; Li, J.; Jabbour, G.E. Excimer-based white phosphorescent organic light-emitting diodes with nearly 100 % internal quantum efficiency. *Adv. Mater.* 2007, 19, 197–202.

[80] Sasabe, H.; Takamatsu, J.-I.; Motoyama, T.; Watanabe, S.; Wagenblast, G.; Langer, N.; Molt, O.; Fuchs, E.; Lennartz, C.; Kido, J. High-efficiency blue and white organic light-emitting devices incorporating a blue iridium carbene complex. *Adv. Mater.* 2010, 22, 5003–5007.

[81] Lindla, F.; Boesing, M.; van Gemmern, P.; Bertram, D., Keiper, D.; Heuken, M.; Kalisch, H.; Jansen, R.H. Employing exciton transfer molecules to increase the lifetime of phosphorescent red organic light emitting diodes. *Appl. Phys. Lett.* 2011, 98, 173304-1–173304-3.

[82] Wang, J.; Liu, J.; Huang, S.; Wu, X.; Shi, X.; He, G.; Zheng, Y. Improved efficiency roll-off at high brightness in simplified phosphorescent organic light emitting diodes with a crossfading-host. *Org. Electron* 2013, 14, 2682–2686.

[83] Jou, J.-H.; Wu, P.-H.; Lin, C.-H.; Wu, M.-H.; Chou, Y.-C.; Wang, H.-C.; Shen, S.-M. Highly efficient orange-red organic light-emitting diode using double emissive layers with stepwise energy-level architecture. *J. Mater. Chem.* 2010, 20, 8464–8466.

[84] Zhou, X.; Qin, D.S.; Pfeiffer, M.; Blochwitz-Nimoth, J.; Werner, A.; Drechsel, J.; Maennig, B.; Leo, K.; Bold, M.; Erk, P.; Hartmann, H. High-efficiency electrophosphorescent organic light-emitting diodes with double light-emitting layers. *Appl. Phys. Lett.* 2002, 81, 4070–4072.

[85] Lee, J.; Lee, J.-I.; Chu, H.Y. Investigation of double emissive layer structures on phosphorescent blue organic light-emitting diodes. *Synth. Met.* 2009, 159, 1460–1463.

[86] Jeon, W.S.; Park, T.J.; Kim, S.Y.; Pode, R.; Jang, J.; Kwon, J.H. Low roll-off efficiency green phosphorescent organic light-emitting devices with simple double emissive layer structure. *Appl. Phys. Lett.* 2008, 93, 063303-1–063303-3.

[87] Liu, Y.; Guo, J.; Feng, J.; Zhang, H.; Li, Y.; Wang, Y. High-performance blue electroluminescent devices based on hydroxyphenyl-pyridine beryllium complex. *Appl. Phys. Lett.* 2001, 78, 2300–2302.

[88] Sun, Y.; Forrest, S.R. High-efficiency white organic light emitting devices with three separate phosphorescent emission layers. *Appl. Phys. Lett.* 2007, 91, 263503-1–263503-3.

[89] Yook, K.S.; Lee, J.Y. High efficiency and low efficiency roll off in white phosphorescent organic light-emitting diodes by managing host structures. *Appl. Phys. Lett.* 2008, 92, 193308-1–193308-3.

[90] Lan, Y.-H.; Hsiao, C.-H.; Lee, P.-Y.; Bai, Y.-C.; Lee, C.-C.; Yang, C.-C.; Leung, M.-K.; Wei, M.-K.; Chiu, T.-L.; Lee, J.-H. Dopant effects in phosphorescent white organic light-emitting device with double-emitting layer. *Org. Electron.* 2011, 12, 756–765.

[91] Bhansali, U.S.; Jia, H.; Oswald, I.W.H.; Omary, M.A.; Gnade, B.E. High efficiency warm-white organic light emitting diodes from a single emitter in graded-doping device architecture. *Appl. Phys. Lett.* 2012, 100, 183305-1–183305-4.

[92] Reineke, S.; Schwartz, G.; Walzer, K.; Leo, K. Reduced efficiency roll-off in phosphorescent organic light emitting diodes by suppression of triplet-triplet annihilation. *Appl. Phys. Lett.* 2007, 91, 123508-1–123508-3.

[93] Zheng, T.; Choy, W.C.H. Red organic light emitting devices with reduced efficiency roll-off behavior by using hybrid fluorescent/phosphorescent emission structure. *Thin Solid Films* 2010, 519, 872–875.

[94] a) Jeon, W.S.; Park, T.J.; Kim, S.Y.; Pode, R.; Jang, J.; Kwon, J.H. Ideal host and guest system in phosphorescent OLEDs. *Org. Electron.* 2009, 10, 240–246 b) Namdas, E.B.; Ruseckas, A.; Samuel, I.D.W.; Lo, S.C.; Burnn P.L. Triplet exciton diffusion in *fac-tris*(2-phenylpyridine) iridium(III)-cored electroluminescent dendrimers. *Appl. Phys. Lett.* 2005, 86, 091104-1–091104-3.

[95] Zheng, T.; Choy, W.C.H.; Ho, C.-L.; Wong, W.-Y. Improving efficiency roll-off in organic light emitting devices with a fluorescence-interlayer-phosphorescence emission architecture. *Appl. Phys. Lett.* 2009, 95, 133304-1–133304-3.

[96] Zheng, T.; Choy, W.C.H. High efficiency blue organic LEDs achieved by an integrated fluorescence–interlayer–phosphorescence emission architecture. *Adv. Funct. Mater.* 2010, 20, 648–655.

[97] Hong, L.-A.; Vu, H.-T.; Juang, F.-S.; Lai, Y.-J.; Yeh, P.-H.; Tsai, Y.-S. Effects of electron blocking and hole trapping of the red guest emitter materials on hybrid white organic light emitting diodes. *Thin Solid Films* 2013, 544, 59–63.

[98] Zhao, F.; Zhang, Z.; Liu, Y.; Dai, Y.; Chen, J.; Ma, D. A hybrid white organic light-emitting diode with stable color and reduced efficiency roll-off by using a bipolar charge carrier switch. *Org. Electron.* 2012, 13, 1049–1055.

[99] Li, W.; Yu, J., Wang, T.; Jiang, Y.; Wei, B. Electroluminescence of organic light-emitting diodes with an ultra-thin layer of dopant. *Mat. Sci. Eng. B* 2008, 149, 77–81.

[100] Yang, X.; Wu, F.-I.; Haverinen, H.; Li, J.; Cheng, C.-H.; Jabbour, G.E. Efficient organic light-emitting devices with platinum-complex emissive layer. *Appl. Phys. Lett.* 2011, 98, 033302-1–033302-3.

[101] a) Su, S.-J.; Gonmori, E.; Sasabe, H.; Kido, J. Highly efficient organic blue-and white-light-emitting devices having a carrier- and exciton-confining structure for reduced efficiency roll-off. *Adv. Mater.* 2008, 20, 4189–4194 b) He, G.; Pfeiffer, M.; Leo, K.; Hofmann, M.; Brinstock, J.; Pudzich, R.; Salbeck, J. High-efficiency and low-voltage *p-i-n* electrophosphorescent organic light-emitting diodes with double-emission layers. *Appl. Phys. Lett.* 2004, 85, 3911–3913.

[102] Yu, J.; Lin, H.; Tong, L.; Li, C.; Zhang, H.; Zhang, J.; Wang, Z.; Wei, B. Enhanced efficiency and reduced roll-off in white organic light-emitting diodes based on two ultra-thin emitting layers. *Phys. Status Solidi A* 2013, 210, 408–412.

[103] Schwartz, G.; Reineke, S.; Walzer, K.; Leo, K. Reduced efficiency roll-off in high-efficiency hybrid white organic light-emitting diodes. *Appl. Phys. Lett.* 2008, 92, 053311-1–053311-3.

[104] Su, Z.; Li, W.; Chu, B.; Wu, S. Triplet to singlet transition induced low efficiency roll-off in green phosphorescent organic light-emitting diodes. *Thin Solid Films* 2011, 519, 2540–2543.

[105] Zhu, Y.-C.; Zhou, L.; Li, H.-Y.; Xu, Q.-L.; Teng, M.-Y.; Zheng, Y.-X.; Zuo, J.-L.; Zhang, H.-J.; You, X.-Z. Highly efficient green and blue-green phosphorescent OLEDs based on iridium complexes with the tetraphenylimidodiphosphinate ligand. *Adv. Mater.* 2011, 23, 4041–4046.

[106] Zhou, G.; Wong, W.-Y.; Yang, X. New design tactics in OLEDs using functionalized 2-phenylpyridine-type cyclometalates of iridium (III) and platinum (II). *Chem. Asian J.* 2011, 6, 1706–1727.

[107] Chen, Z.Q.; Bian, Z.Q.; Huang, C.H. Functional Ir^{III} complexes and their applications. *Adv. Mater.* 2010, 22, 1534–1539.

[108] Li, M.; Wang, Q.; Dai, J.; Lu, Z.-Y.; Huang, Y.; Yu, J.-S.; Luo, S.; Su, S.-JHighly efficient orange and red organic light-emitting diodes with iridium (III) complexes bearing benzothiazole cyclometallate ligands as emitters. *Thin Solid Films* 2012, 526, 231–236.

[109] Li, X.; Zang, T.; Yu, L.; Zhang, D.; Lu, G.; Chi, H.; Xiao, G.; Dong, Y.; Cui, Z.; Zhang, Z.; Hu, Z. A promising phosphorescent heteroleptic iridium complex with carbazole-functionalized substituent: Synthesis, photophysical and electroluminescent performances. *Opt. Mater.* 2012, 35, 300–306.

[110] a) Ono, K.; Joho, M.; Saito, K.; Tomura, M.; Matsushita, Y.; Naka, S.; Okada, H.; Onnagawa, H. Synthesis and electroluminescence properties of *fac-tris*(2-phenylpyridine)iridium derivatives containing hole-trapping moieties. *Eur. J. Inorg. Chem.* 2006, 3676–3683 b) Liu, Z.; Nie, D.; Bian, Z.; Chen, F.; Lou, B.; Bian, J.; Huang, C. Photophysical properties of heteroleptic iridium complexes containing carbazole-functionalized β-diketonates. *Chem. Phys. Chem.* 2008, 9, 634–640 c) Wong, W.-Y.; Ho, C.-L.; Gao, Z.-Q.; Mi, B.-X.; Chen, C.-H.; Cheah, K.-W.; Lin, Z. Multifunctional iridium complexes based on carbazole modules as highly efficient electrophosphors. *Angew Chem. Int. Ed.* 2006, 45, 7800–7803 d) Ho, C.-L.; Wang, Q.; Lam, C.-S.; Wong, W.-Y.; Ma, D.; Wang, L.; Gao, Z.-Q.; Chen, C.-H.; Cheah, K.-W.; Lin, Z. Phosphorescence color tuning by ligand, and substituent effects of multifunctional iridium(III) cyclometalates with 9-arylcarbazole moieties. *Chem. Asian J.* 2009, 4, 89–103 e) Ding, J.; Gao, J.; Cheng, Y.; Xie, Z.; Wang, L.; Ma, D.; Jing, X.; Wang, F. Highly efficient green-emitting phosphorescent iridium dendrimers based on

carbazole dendrons. *Adv. Funct. Mater.* 2006, 16, 575–581 f) Ho, C.-L.; Wong, W.-Y.; Yao, B.; Xie, Z.; Wang, L.; Lin, Z. Synthesis, characterization, photophysics and electrophosphorescent applications of phosphorescent platinum cyclometalated complexes with 9-arylcarbazole moieties. *J. Organomet. Chem.* 2009, 694, 2735–2749 g) Wang, Q.; Ho, C.-L.; Zhao, Y.; Ma, D.; Wong, W.-Y.; Wang, L. Reduced efficiency roll-off in highly efficient and color-stable hybrid WOLEDs: The influence of triplet transfer and charge-transport behavior on enhancing device performance. *Org. Electron.* 2010, 11, 238–246 h) Ho, C.-L.; Chi, L.-C.; Hung, W.-Y.; Chen, W.-J.; Lin, Y.-C.; Wu, H.; Mondal, E.; Zhou, G.-J.; Wong, K.-T.; Wong, W.-Y. Carbazole-based coplanar molecule (CmInF) as a universal host for multi-color electrophosphorescent devices. *J. Mater. Chem.* 2012, 22, 215–224 i) Ho, C.-L.; Wong, W.-Y.; Zhou, G.-J.; Yao, B.; Xie, Z.; Wang, L. Solution-processible multi-component cyclometalated iridium phosphors for high-efficiency orange-emitting OLEDs and their potential use as white light sources. *Adv. Funct. Mater.* 2007, 17, 2925–2936 j) Ho, C.-L.; Wong, W.-Y.; Gao, Z.-Q.; Chen, C.-H.; Cheah, K.-W.; Yao, B.; Xie, Z.-Y.; Wang, Q.; Ma, D.-G.; Wang, L.-X.; Yu, X.-M.; Kwok, H.-S.; Lin, Z.-Y. Red-light-emitting iridium complexes with hole-transporting 9-arylcarbazole moieties for electrophosphorescence efficiency/color purity trade-off optimization. *Adv. Funct. Mater.* 2008, 18, 319–331 k) Al-Attar, H.A.; Griffiths, G.C.; Moore, T.N.; Tavasli, M.; Fox, M.A.; Bryce, M.R.; Monkman, A.P. Highly efficient, solution-processed, single-layer, electrophosphorescent diodes and the effect of molecular dipole moment. *Adv. Funct. Mater.* 2011, 21, 2376–2382 l) Tavasli, M.; Moore, T.N.; Zheng, Y.; Bryce, M.R.; Fox, M.A.; Griffiths, G.C.; Jankus, V.; Al-Attar, H.A.; Monkman, A.P. Colour tuning from green to red by substituent effects in phosphorescent *tris*-cyclometalated iridium (III) complexes of carbazole-based ligands: synthetic, photophysical, computational and high efficiency OLED studies. *J. Mater. Chem.* 2012, 22, 6419–6428 m) Shan, G.-G.; Li, H.-B.; Mu, Z.-C.; Zhu, D.-X.; Su, Z.-M.; Liao, Y. Synthesis of cationic iridium(III) complexes with high quantum yield via enhancing the steric hindrance of ligands. *J. Organomet. Chem.* 2012, 702, 27–35.

[111] a) Eng, T.; Li, G.; Ye, K.; Huang, S.; Wu, Y.; Liu, Y.; Wang, Y. Concentration-insensitive and low-driving-voltage OLEDs with high efficiency and little efficiency roll-off using a bipolar phosphorescent emitter. *Org. Electron.* 2013, 14, 1649–1655 b) Peng, T.; Li, G.; Ye, K.; Wang, C.; Zhao, S.; Liu, Y.; Hou, Z.; Wang, Y. Highly efficient phosphorescent OLEDs with host-independent and concentration-insensitive properties based on a bipolar iridium complex. *J. Mater. Chem. C* 2013, 1, 2920–2926.

[112] a) Wang, Q.; Yu, J.; Li, M.; Lu, Z. Tailoring the spectra of white trganic light-emitting devices by trap effect of a concentration-insensitive dopant. *Int. J. Photoenergy* 2013, 392408-1–392408-6 b) Dumur, F.; Lepeltier, M.; Zamani Siboni, H.; Xiao, P.; Graff, B.; Lalevée, J.; Gigmes, D.; Aziz, H. Concentration-insensitive phosphorescent organic light emitting devices (PhOLEDs) for easy manufacturing. *J. Lumin.* 2014, 151, 34–40.

[113] Cai, J.-X.; Ye, T.-L.; Fan, X.-F.; Han, C.-M.; Xu, H.; Wang, L.-L.; Ma, D.-G.; Lin, Y.; Yan, P.-F. An effective strategy for small molecular solution-processable iridium (III) complexes with ambipolar characteristics: towards efficient electrophosphorescence and reduced efficiency roll-off. *J. Mater. Chem.* 2011, 21, 15405–15416.

[114] Wang, J.; Yu, J.; Li, L.; Wang, T.; Yuan, K.; Jiang, Y. Low roll-off power efficiency organic light-emitting diodes consisted of nondoped ultrathin phosphorescent layer. *Appl. Phys. Lett.* 2008, 92, 133308-1–133308-3.

[115] a) Choukri, H.; Fischer, A.; Forget, S.; Chénais, S.; Castex, M.; Adès, D.; Siove, A.; Geffroy, B. White organic light-emitting diodes with fine chromaticity tuning via ultrathin layer position shifting. *Appl. Phys. Lett.* 2006, 89, 183513-1–183513-3 b) Tsuji, T.; Naka, S.; Okada, H.; Onnagawa, H. Nondoped-type white organic electroluminescent devices utilizing complementary color and exciton diffusion. *Appl. Phys. Lett.* 2002, 81, 3329–3331.

[116] Lei, X.; Yu, J.; Zhao, J.; Jiang, Y. Electroluminescence of organic light-emitting diodes consisting of an undoped $(pbi)_2Ir(acac)$ phosphorescent layer. *Physica B* 2011, 406, 4249–4252.

[117] Zhu, M.; Zou, J.; He, X.; Yang, C.; Wu, H.; Zhong, C.; Qin, J.; Cao, Y. Triphenylamine dendronized iridium (III) complexes: robust synthesis, highly efficient nondoped orange electrophosphorescence and the structure–property relationship. *Chem. Mater.* 2012, 24, 174–180.

[118] Wang, K.; Zhao, F.; Wang, C.; Chen, S.; Chen, D.; Zhang, H.; Liu, Y.; Ma, D.; Wang, Y. High-performance red, green, and blue electroluminescent devices based on blue emitters with small singlet–triplet splitting and ambipolar transport property. *Adv. Funct. Mater.* 2013, 23, 2672–2680.

[119] Jou, J.-H.; Yu, H.-H.; Lin, Y.-X.; Tseng, J.-R.; Peng, S.-H.; Jou, Y.-C.; Lin, C.-H.; Shen, S.-M.; Hsieh, C.-Y.; Wei, M.-K.; Lin, D.-H.; Wang, C.-C.; Chen, C.-C.; Tung, F.-C.; Chen, S.-H.; Wang, Y.-S. High efficiency yellow organic light emitting diodes with a balanced carrier injection co-host structure. *J. Mater. Chem. C* 2013, 1, 5110–5115.

[120] Kanno, H.; Sun, Y.; Forrest, S.R. White organic light-emitting device based on a compound fluorescent-phosphor-sensitized-fluorescent emission layer. *Appl. Phys. Lett.* 2006, 89, 143516-1–143516-3.

[121] a) Sun, Y.; Giebink, N.C.; Kanno, H.; Ma, B.; Thompson, M.E.; Forrest, S.R. Management of singlet and triplet excitons for efficient white organic light-emitting devices. *Nature* 2006, 440, 908–912 b) Sun, Y.; Giebink, N.C.; Forrest, S.R. Transient analysis of triplet exciton dynamics in amorphous organic semiconductor thin films. *Org. Electron.* 2006, 7, 375–386.

[122] a) Schwartz, G.; Pfeiffer, M.; Reineke, S.; Walzer, K.; Leo, K. Harvesting triplet excitons from fluorescent blue emitters in white organic light-emitting diodes. *Adv. Mater.* 2007, 19, 3672–3676 b) Zhang, Y.; Forrest, S.R. Triplets contribute to both an increase and loss in fluorescent yield in organic light emitting diodes. *Phys. Rev. Lett.* 2012, 108, 267404-1–267404-5.

[123] Meng, L.-C.; Hou, Y.-B.; Lou, Z.-D.; Teng, F.; Yao, X.; Liu, X.-J.; Tang, A.-W.; Peng, J.-B. Efficient and color-stable white organic light-emitting diodes based on exciton management and phosphorescent sensitization. *Synth Met.* 2013, 172, 63–68.

[124] a) Baek, H.; Lee, C. Simple white organic light emitting diodes with improved color stability and efficiency using phosphorescent and fluorescent emitters. *J. Appl. Phys.* 2008, 103, 124504-1–124504-5 b) Yang, S.-H.; Liu, M.-H.; Su, Y.-K. Stable and highly bright white organic light-emitting diode based on 4,4',4''-*tris*(*N*-3-methylphenyl-*N*-phenyl-amino)triphenylamine. *J. Appl. Phys.* 2006, 100, 083111-1–083111-4 c) Hsiao,

C.-H.; Lin, C.-F.; Lee, J.-H. Driving voltage reduction in white organic light-emitting devices from selectively doping in ambipolar blue-emitting layer. *J. Appl. Phys.* 2007, 102, 094508-1–094508-6 d) Kim, C.-H.; Shinar, J. Bright small molecular white organic light-emitting devices with two emission zones. *Appl. Phys. Lett.* 2002, 80, 2201–2203 e) Krummacher, B.C.; Choong, V.-E.; Mathai, M.K.; Choulis, S.A.; So, F.; Jermann, F.; Fiedler, T.; Zachau, M. Highly efficient white organic light-emitting diode. *Appl. Phys. Lett.* 2006, 88, 113506-1–113506-3 f) Ho, M.-H.; Hsu, S.-F.; Ma, J.-W.; Hwang, S.-W.; Yeh, P.-C.; Chen, C.H. White *p-i-n* organic light-emitting devices with high power efficiency and stable color. *Appl. Phys. Lett.* 2007, 91, 113518-1–113518-3 g) Jou, J.H.; Chiu, Y.S.; Wang, R.Y.; Hu, H.C.; Wang, C.P.; Lin, H.W. Efficient, color-stable fluorescent white organic light-emitting diodes with an effective exciton-confining device architecture. *Org. Electron.* 2006, 7, 8–15 h) Jou, J.H.; Wu, M.H.; Wang, C.P.; Chiu, Y.S.; Chiang, P.H.; Hu, H.C.; Wang, R.Y. Efficient fluorescent white organic light-emitting diodes using co-host/emitter dual-role possessed di(triphenyl-amine)-1,4-divinyl-naphthalene. *Org. Electron.* 2007, 8, 735–742 i) Chang, C.-C.; Chen, J.-F.; Hwang, S.-W.; Chen, C.H. Highly efficient white organic electroluminescent devices based on tandem architecture. *Appl. Phys. Lett.* 2005, 87, 253501-1–253501-3 j) Guo, F.W.; Ma, D.G. White organic light-emitting diodes based on tandem structures. *Appl. Phys. Lett.* 2005, 87, 173510-1–173510-3 k) Lai, S.L.; Chan, M.Y.; Fung, M.K.; Lee, C.S.; Lee, S.T. Copper hexadecafluoro-phthalocyanine and copper phthalocyanine as a pure organic connecting unit in blue tandem organic light-emitting devices. *J. Appl. Phys.* 2007, 101, 014509-1–014509-4 l) Ho, M.-H.; Chen, T.-M.; Yeh, P.-C.; Hwang, S.-W.; Chen, C.H. Highly efficient *p-i-n* white organic light emitting devices with tandem structure. *Appl. Phys. Lett.* 2007, 91, 233507-1–233507-3 m) Lee, T.-W.; Noh, T.; Choi, B.-K.; Kim, M.-S.; Shin, D.W.; Kido, J. High-efficiency stacked white organic light-emitting diodes. *Appl. Phys. Lett.* 2008, 92, 043301-1–043301-3 n) Yeh, S.J.; Chen, H.Y.; Wu, M.F.; Chan, L.H.; Chiang, C.L.; Yeh, H.C.; Chen, C.T.; Lee, J.H. All non-dopant red–green–blue composing white organic light-emitting diodes. *Org. Electron.* 2006, 7, 137–143.

[125] Yook, K.S.; Jeon, S.O.; Joo, C.W.; Lee, J.Y. High efficiency, color stability, and stable efficiency roll off in three color hybrid white organic light emitting diodes. *Appl. Phys. Lett.* 2008, 93, 073302-1–073302-3.

[126] Najafabadi, E.; Knauer, K.A.; Haske, W.; Kippelen, B. High-performance inverted top-emitting green electrophosphorescent organic light-emitting diodes with a modified top Ag anode. *Org. Electron.* 2013, 14, 1271–1275.

INDEX

D

E

F

G

H

I

P

Q

R

S

T

U

V

W

X

Y

Z